The *Good Flight Simmer's Guide*

Edited by Mike Clark

The Essential Guide for the PC Aviator

2002

Published by TecPilot

TecPilot Publishing,
76 Victoria Road,
Parkstone, Poole,
Dorset, BH12 3AF,
United Kingdom

First Published 2001

© TecPilot & Mike Clark, 2001

ISBN 0 9541967 0 8

Cover designed by Mike Clark

Printed and bound in the United Kingdom

To my wife Lesley who put up with my frustration during the making of this book. You are my rock, my friend, my life, my angel.

To my cats; Sapphire, Milly, Poppy, Fluff, Jack and Jill, my beautiful, unconditional loving friends who remind me that only the important things in life matter.

DEPARTURES

To Ray F Jones who passed away during the making of this book. He was an inspiration, a solid friend and one who gave selflessly to this wonderfully complex hobby. He is still missed.

To Andreas, whose life was suddenly taken in a tragic road accident during the making of the book. He flew in life but now glides in spirit.

Credits & Acknowledgements

Sincere thanks go to the following people and organisations who assisted during the making of this book. I am extremely grateful for your help.

CONTRIBUTORS

Lesley Clark - Sub Editing
Evan Hochberg - Product listings *
Greg Gott - Hardware for flight simulation
Jos Grupping - Flight Simulator history - http://www.simflight.com/fshistory
Ray F Jones - Installing downloads into Flight Simulator
Rich Kaminsky - Aviation Museums
Stephen Heyworth - Advanced Flying - The Flight of your life

All other content and artwork by - **Mike Clark**

INFORMATION, ADVICE & INSPIRATION

Adobe - http://www.adobe.com
Andy Payne - The Producers - http://www.the-producers.co.uk
AOPA (UK) - Aircraft Owners and Pilots Association - http://www.aopa.co.uk
Arnie Lee - Abacus Publishing - http://www.abacuspub.com
Bob Sidwick - RC Simulations - http://www.rcsimulations.com
Enrico Shiratti - Project Magenta - http://www.shiratti.com
Gene & Mary - OneMileUp Inc. http://www.onemileup.com
Hewlett Packard - http://www.hp.com
Katy Pluta - The SimFlight Network - http://www.simflight.com
Jos Grupping - Flight Simulator History - http://www.simflight.com/fshistory
Judy Smith - TecPilot Corporate Projects
Mathijs Kok - Lago snc http://www.lagoonline.com
Microsoft (USA) - http://www.microsoft.com
Microsoft (UK) - http://www.microsoft.co.uk
Miguel Blaufuks - The SimFlight Network - http://www.simflight.com
Mike Bannister - British Airways - http://ww.britishairways.com
Mungo Amyatt-Leir - Just Flight Ltd - http://www.justlfight.com
Nels Anderson - FlightSim.com - http://www.flightsim.com
Paul Sunter - B&L Production Ltd - http://www.bl-distribution.co.uk
Richard Branson - Virgin - http://www.virgin.com
Rob Jubb - TecPilot News Editor
Robert Kirkland - Phoenix Simulations - http://www.phoenix-simulation.co.uk
Robert Stallibrass - Contact Sales http://www.contact-sales.co.uk
Tom Allensworth - Avsim Online - http://www.avsim.com
Trevor Morson - The DC3 Hangar - http://www.douglasdc3.com
Winfried Diekmann - Aerosoft - http://www.aerosoft.com
Yahoo! - http://www.yahoo.com

* A special thanks to Evan Hochberg for what must have seemed an endless job building the product listings from scratch. Thanks for your patience and endless dedication.

Contents

Contents

The Good Flight Simmer's Guide 2002

Welcome to *The Good Flight Simmer's Guide 2002*. Everything you need know about "flying" Flight Simulator as a virtual aviator instead of "playing" it as a gamer.

The first question that most people ask after using Microsoft Flight Simulator is, "what's next?". In this book we answer that question and give you the insight into a hobby that goes far beyond playing a computer "game". This is NOT a tips, tricks or cheats book, but a guide take you to the heart of making aviation "As Real As It Gets" on a personal computer.

Initially we'll give you an idea of how far flight simulation has flourished since its conception in 1979 to the present day in; *The History of Flight Simulator*. A title that has sold consistently over the past 20 years and is recognised as THE most popular simulation title world-wide. Over 21 Million copies have been sold to-date, according to The Guinness Book of Records.

Then, we shall demonstrate how to get the most out of your computer by providing vital information about setting up your hardware specifically for use with flight simulators.

We'll investigate why the Internet plays an important roll in the hobby, discover how you can get the most out of the web, and save you money at the same time by providing links to where you can get FS software for FREE.

Once you have read the Internet section, we guarantee you will want to download a number of files. So in Section Four we felt it prudent to explain, in a simple language, just how to install them properly and optimise them for use.

After all that downloading and installing, no doubt you will need to go flying. We'll take you on two extremely adventurous flights; VFR (Visual Flight Rules) and IFR (Instrument Flight Rules). We'll give you an idea of how realistic flight simming can be. Don't think for a moment that it will be a walk in the park because in Section Six we'll take you on the "*Flight of your Life!*" Fly around the British Isles attempting 5 complex approaches and landings. If you think that's a breeze then we'll have to keep you busy by asking one brain busting question for every week of the year in the *52 Week Aviation Quiz!* Don't worry, for those of you with an insatiable curiosity we have provided answers at the back of the book. If you love flight sim so much that you simply cannot get enough of it we've even included a crossword and word search for you to ponder over on the way to work.

Most flight simmers are bewildered by the volume of add-on software available for Flight Simulator today. Some get confused about compatibility issues while others don't realize that there are hundreds of products choose from in the shops or the Internet. Do not despair! For the first time EVER we have assembled a COMPLETE listing of add-on products spanning 7 simulation titles, over 4 years, and stuffed them into this book. Amazingly, that's over 300 products. Some of them are current while others are not. We felt it only fair that we should included every single title of note to give you the virtual pilot, developer or publisher a chance to see what's been published and what potential there is for the future. You could say we are attempting to be inspirational too.

If that isn't enough to wet your whistle then we have included a full list of web sites in aviation and simulation and topped it off with an abundance of reference material to-boot. Be assured if it's not in this book we are working on it for the next version of *The Good Flight Simmer's Guide*.

Good reading and good flying!

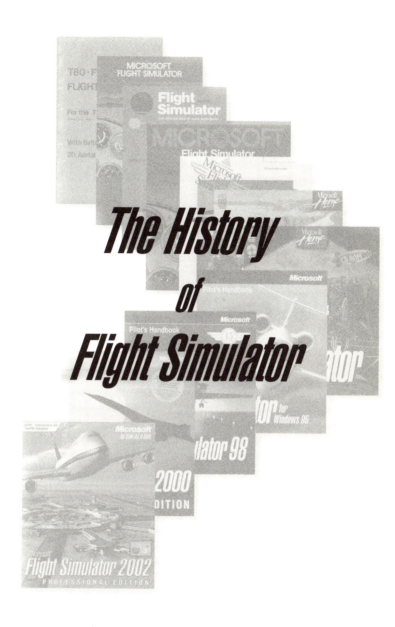

The History of Flight Simulator

The History of Flight Simulator

In October 2001 the latest version of Flight Simulator was launched by Microsoft - Flight Simulator 2002 (FS2002). Hundreds of enthusiasts and indeed pilots were involved in its development, programming and beta testing. It would appear that millions of copies have already sold worldwide of what is commonly considered as the "Eighth Generation" of this hugely successful franchise.

So if this is the Eighth generation, what about the first and consecutive versions and what did they look like? In this first section we thought we should give you a complete overview of the history of Flight Simulator since its first release in 1979. Since its conception many different versions have been released and for a variety of operating systems and machines. In fact, Flight Simulator has become a catalyst for one of the biggest aviation genres in the world. Thanks to the inventiveness and dedication of a student called Bruce Artwick.

In the mid-70's Bruce Artwick was an electrical engineering student at the University of Illinois. Being a passionate pilot, it was only natural that the principles of flight became the focus of his studies. In his thesis of May 1975, called '*A versatile computer-generated dynamic flight display*', he presented the flight model of an aircraft displayed on a computer screen. He proved that a 6800 processor (the first available microcomputer at the time) was able to handle both the arithmetic and the graphic display needed for real-time flight simulation. In short: the first flight simulator was born.

The Birth of Flight Simulator...

In 1978 Bruce Artwick, together with Stu Moment, founded a software company by the name of subLOGIC and started developing graphics programs for the 6800, 6502, 8080 and other processors. In 1979 Bruce decided to take the model from his original thesis one step further and developed the first flight simulator program for the Apple-II (based on the 6502 processor). This followed a version ported for the Radio Shack TRS-80. Both versions were constructed in their respective machine-code and both were loaded from a cassette tape!

Apple II Computer

subLOGIC FS1 hit the consumer market in November 1979 (see advert). Flight Simulator earned a reputation as being the best selling title for the Apple in 1981. By the end of 1997 Microsoft claimed that all their versions had sold not less than 10 million copies, making it the best selling software title in the 'entertainment' sector. In the year 2000 The Guinness Book of Records confirmed this by reporting that as of June 1999 Flight Simulator had sold a massive 21 million copies worldwide.

The screenshot (right) is from one of the first releases for the Apple II. As you can see it shows a resemblance to subsequent versions. At the bottom of the screen was a 'panel' (albeit simple) with a selection of gauges. On top was a three-dimensional (wire frame) display that rendered the scenery from a pilots perspective. Over 23 aircraft characteristics were taken into account and the frame rate would have been around 3-6 frames per second (A good frame rate today would be 25 fps - 60 fps). The terrain was small and flat with rivers running, or should we say drawn, across the landscape. It even had a 3 airports! Objects like mountains and bridges were already starting to make an appearance at this early stage too.

Microsoft Flight Simulator 1.00

Bruce Artwick's work didn't go unnoticed. Another clever young man from Redmond, USA had just set up his own small software company called Microsoft and was shifting his attention from the Commodore 64 (C64) to the newly developed IBM-PC. This gentleman went by the name of Bill Gates who soon set about entered a bidding war with IBM to obtain a license for Flight Simulator. Which he achieved.

Microsoft Flight Sim 1.00

In November 1982 **Microsoft Flight Simulator 1.00** hit the stores as one of the first PC entertainment titles, followed by version 2.00 a year later. More releases followed until version 2.14 in 1986. Although the CGA card and RGB monitor could only sustain 4 colours, a clever dithering (colour-mixing) system was added that magically generated an extra 6 colours. This helped vastly in the rendering of scenery and at the time made it look quite real. The focus on aircraft at that time was the Cessna 182 that even included a retractable landing gear!

Microsoft Flight Sim 1.00

All the necessary flight controls were included in the panel and the minimum VFR (Visual Flight Rule) and IFR (Instrument Flight Rule) and associated communications equipment as required by the FAA (Federal Aviation Administration) were there too. OBI, ILS and DME were also present. The only important instrument that was missing was an ADF (Automatic Direction Finder). The use of a joystick was possible and even recommended! Amazing!

MS-FS also featured a new and sophisticated co-ordinate system for the flight sim world developed by Bruce Artwick. In the new scheme the "world" had a flat surface of 10.000 x 10,000 square miles with a basic resolution of around 2.5 inches. The area encompassed all of the Continental USA, extending into Canada, Mexico and the Caribbean. The "populated" world consisted of 4 small areas: Chicago, Seattle, Los Angeles and New York/Boston. A total of 20 airports occupied the co-ordinated area which increased to 80 in later versions. In-between was "nothing-land" that contained no scenery, no airports or radio-stations. You didn't want to go there on a dark night!

Microsoft Flight Sim 1.02 for Mac

The number of aircraft characteristics had been raised to 35. When looking from a distance this version bears an even more remarkable resemblance to versions you see today. But it still wasn't possible to see cockpit interior or exteriors. You had to take Artwick's and Microsoft's word that it was indeed a Cessna you were flying. It was quite an achievement that all this worked on a 64K IBM-PC loaded from a single 5¼" floppy disk. Compared to FS-1 for the Apple II, and TRS-80, this version could truly be named as a "Second Generation" flight simulator.

Microsoft Flight Simulator 2.00

Flight Simulator II

From here on up until and including Microsoft Flight Simulator 3.0 the version history is a little vague. However, in the interim period Microsoft and subLOGIC both released two separate versions but for different computers. subLOGIC released Flight Simulator II for the Apple II (1983) while MS presented Flight Simulator 2 for the IBM PC.

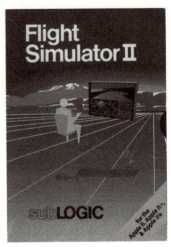

subLOGIC's boasted a superior colour display which was made possible utilising the Apple's massively "fast" 48K system. The aircraft was a Piper Cherokee Archer and scenery had the same 4 areas as Microsoft's which incorporated over 80 airports! The ADF (Automatic Direction Finder) was an extra. Later, similar versions were ported to the Commodore 64 and Atari 800 and XL. All these versions are now considered as the "second generation" of the development of the overall title. This time however, users could now enjoy much better graphics! In 1987 Microsoft moved another step forward and published a refined edition for the first generation "Apple MacIntosh". This was however in monochrome only.

Between 1984 and 1987 subLOGIC released at least 14 versions of Flight Simulator II for a diverse range of personal computers, including the Commodore Amiga and Atari ST. A rare variant was published for the Philips MSX-computer which possessed the unexpected and strange bonus of "*Torpedo Attack*".

Later releases of FSII for the Amiga, Atari ST and MacIntosh were in reality a completely new concept. Not only were the number of calculations raised to 49 but the resolution was promoted to 1/100 of an inch. There was also a complete new interface with a menu bar at the top and dropdown submenus, like in modern applications instead of the previously awkward editor. To add a bit of spice the Piper was replaced by a Cessna 182 together with a representation of a Learjet 25G. Sadly, the flight characteristics and bog standard panel (used for both aircraft) weren't very realistic. However, with the addition of the all new and shiny "San Francisco Scenery" the number of areas were increased to 5, making a total of 120 airports. This seemed to make up for other minor disappointments.

The superior quality of the 68000 processor made for a stunning display across all of the aforementioned versions. Image quality was vastly improved with hidden surface elimination and surface shadowing. Day, night and cloud effects were more realistic and, for the first time, it was possible to see ones own aircraft from the outside - both in spot <u>and</u> tower views. A separate movable map was thrown in for good measure and all these

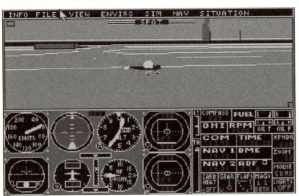

new views included zoom features as well. Pilots could even monitor their mistakes with an "instant replay" feature - which is standard today.

Lastly, a Multi-Player option was added, which made it possible for two pilots to fly together via networked computers. Soon after this facility was discovered clubs like FSFAN in Europe extended the capabilities over multiple networks. This enabled entire groups of pilots to fly together in real-time. Flight Simulation was now motoring and simmers had never had it so good!

The first add-on scenery disks

Until 1987 the "flyable" world in Flight Simulator (airports, navigation aids and some scenery), was restricted to 4 later 5 small sectors around major urban areas. In 1987 this was vastly expanded with the addition of the first series of scenery disks by subLOGIC. Twelve 5¼" scenery disks were developed covering the entire continental US. This was possible by taking advantage of official NOAA (National Oceanic and Atmospheric Administration) Aeronautical Charts and Airport/Facility Directories. The scenery disks also included enough radio-navigation aids as well as visual scenery to allow a user to navigate anywhere in the areas covered. A typical disk included approximately 100 airports and 100 radio-navaids, making them ideal for cross-country flights. VFR (visual flight rules) was starting to take off!

European Flight Simulator pilots had always felt left out up until this point but this was suddenly changed in 1988 when they were furnished with a release by subLOGIC of the *Western European Tour* scenery. In this the whole flat co-ordinate area was relocated to Europe, stretching from Iceland to Luxor in Egypt and far beyond Russia! Detailed scenery was however limited to the southern UK, Northern France and south west Germany and only contained big cities, rivers, roads and coastlines. For the first time though virtual pilots could now fly over the Thames, under Tower Bridge, visit Stonehenge, cross the English Channel, gaze at the Eiffel tower in Paris and even visit the Olympic Stadium in Munich. Another disk was released for Japan as well.

Microsoft Flight Simulator 3.00

Early 1988 Bruce Artwick split with Stu Moment, left subLOGIC and started his own company called BAO Ltd (Bruce Artwick Organisation). It was created for the purpose of developing and marketing simulation products, tailored to Microsoft Flight Simulator. Later that year Microsoft FS 3.0 was released, which featured 16 bit colour graphics at an (EGA) resolution of 640 x 350, a

Microsoft Flight Simulator 3.00 (BAO)

menu system, separate windows and, for the first time, (for Microsoft) exterior views of aircraft. Previous releases of Flight Simulator II (1986/87) for the Atari ST, Amiga and Microsoft FS 1.0 for the MacIntosh already incorporated all of these features.

FS 3.0 included flight lessons, crash analysis, crop dusting and other scenarios. It also had autopilot functionality. If a hard disk was available, multiple scenery libraries could be set up with automatic coordinate setting when switching areas. FS 3.0 was also compatible with the earlier scenery disks by subLOGIC.

Microsoft Flight Simulator 4.00

Microsoft Flight Simulator 4.00

In 1989 Flight Simulator 3.0 for the PC was followed by a similar, but improved **version 4.0**. But was it still possible to run version 3.0 on a 10MHz 8086 PC-XT with 256 Kb memory and a floppy disk? This latest version needed at least a 12 MHz 80286 PC-AT, preferably a 33 MHz 80386 with 384 Kb memory and a hard disk. The basic resolution and number of colours were the same as with FS 3.0. Even the WW1 game with the Sopwith Camel from Flight Simulator 1 was still present with an enemy field and cardboard looking mountains.

First, Flight Simulator 4.0 contained many bug fixes and improved flight characteristics, but also included some new features never seen before. The scenery was greatly enhanced. Runway approach lighting was added with flashing strobes which were particularly spectacular at night. Dynamic scenery was added in the form of tow trucks, taxiing, departing and landing aircraft, balloons and sailboats.

The aircraft had night-lighting and the scenery library now included a Schweitzer 2-32 sailplane. Air traffic controllers provided takeoff and landing clearance (through text instruction). A weather generator could display dynamic weather: clouds, wind and turbulence. Finally, FS 4.0 included a basic aircraft editor that allowed the user to build his own simple aircraft designs.

In 1991 Microsoft released a version 4.0 for the Apple MacIntosh. In many respects it was more advanced than its PC counterpart - especially with respect to graphics and modularity. This was due to the superior quality

Interior of Microsoft Flight Simulator 4.00

and stability of its multi-window, graphics oriented, operating system, as opposed to the two dimensional MS-DOS based platform of the time. The Mac version already featured a superior menu system and un-dockable windows. However, in aviation and simulation terms it was not that much different.

Opening up Flight Simulator

A whole new era started in 1990 when Microsoft, for the first time, made an opening in an historically hermetically sealed product. Until 1990 the only add-ons available were scenery disks from subLOGIC. This radically changed when Microsoft released "Aircraft and Scenery Designer" (A&SD), created by our friends at BAO. This for the first time allowed users to generate their own scenery and aircraft.

Flight Simulator 4.00 for the MacIntosh

A&SD came with some ready made aircraft, including a Beechcraft Starship and the first Boeing 747-400 with a "glass cockpit". More importantly the design process was a simple affair. This was later to change however with the release of BAO's "FlightShop" as Flight Simulator became more and more complex.

The scenery design program was quite powerful. Furthermore, A&SD spawned a wealth of other programs like **SEE** (Scenery Enhancement Editor) by Kikiware (Laemming Wheeler), and **SGA** (Sound, Graphics and Aircraft Update) that contained the first Concorde, **AAF** (Aircraft and Adventure Factory) and **AFD** (Airport and Facility Directory) - all published by Mallard.

The knock on effects were broadened further by many flight simmers trying to unravel the inner secrets of Flight Simulator. All were eager to cash in on the boom in add-on development. However, most were concentrating on the structure of BGL files: Files that govern the basic scenery structure. BGL's soon became referred to as "Bruce's Graphic Language". (see "Installing Scenery for Flight Simulator" for a more complex description)

People like Gavioli, Kukushkin, Borgsteede, Schiratti and many others opened up the structure further by writing programs for "easy" scenery creation. One of the factors that greatly facilitated this process was the emergence of BBS (Bulletin Board System) and email.

The Fifth Generation

Several years went by with no new versions of the franchise appearing on the shelves. Fans started to resign themselves to the fact that Flight Simulator was to be no more. However, their fears were diminished when in 1993 (4 years later), version 5.0 was launched by the BAO/Microsoft partnership - and boy what a release! Bruce Artwick was quoted in an interview as saying something like; "Odd versions (generations) contain new features and techniques. Even versions are refinements". If the former statement was true then it certainly became apparent in Microsoft Flight Simulator 5.0. In fact, Microsoft were so happy with their release they even gave it the strap line "As Real As It Gets".

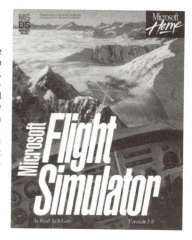

New scenery based upon a worldwide spherical co-ordinate system and a host of other new features blasted their way out of the box and this code was to set a standard for all subsequent versions. However, this new edition upset some users as it outmoded some of Flight Sim 4.00 features. Plus it was now only to be available for the **P**ersonal **C**omputer market!

Flight Simulator 5.0 came accompanied with a very good 284 page Pilot's Handbook and needed a 386, prefera- bly a 486 PC with 530 Kb free memory to run it. The 640x400 display (256 colours) further required an SVGA graphics card. The sound was much improved and users did not have to put up with a nasty hum through their speakers. Digitally sampled engine and other sounds were now the latest fad and made for great listening. The Learjet had been upgraded to a model 35A but users had to put up with a default Cessna panel to fly it. To the delight of many however this "panel" was now referred to as "photorealistic". This was a bit of an overstatement actually but it looked much better nonetheless. Flight models were greatly enhanced too. The autopilot was expanded with an A.O. ILS-lock and a "Land Me" function was built in to help pilots who had fallen asleep at the controls.

The big changes however were in the scenery. Not only was it possible to fly everywhere in the world, due to the new co-ordinate system, but the scenery was now covered with photo realistic textures. This made it look much improved. Due to the increase to 256 colours new dawn/dusk effects were possible and the horizon now blended in gradient type colours. In urban areas many new "cyber graphic" buildings were added with shadow and night lighting effects too which added greatly to the perception of depth.

Unfortunately, the first release contained so many bugs and inconsistencies, that customers were very disap- pointed and lead to many shouting matches on the internet Bulletin Boards and Forums. Microsoft listened however and responded fairly quickly by releasing a bug fix (version 5.0a) in February 1994 that addressed all the problems submitted to them.

Later in 1994 Flight Simulator became more interesting for Europeans, when BAO released a scenery expansion called Europe-1. This covered Germany, Austria, Switzerland and The Netherlands. Europe-2 and 3 followed covering most of Europe with what was then considered highly realistic scenery. A wealth of other companies then joined the bandwagon by providing scenery for other parts of the world. The virtual world started to flourish and even more so with freeware scenery designers now cropping up all over the place.

In the meantime a new fangled medium called CD-ROM(?) appeared on the scene and in 1995 Microsoft took advantage of by releasing a new version Flight Simulator - version 5.1. In reality it was a beefed up version of 5.0 but with a bigger and more enhanced scenery engine. In fact, the scenery engine had been enhanced to 32bit and included better looking textures. A cache system was also incorporated for faster and smoother loading of textures. Improved coastlines, night effects, clouds, and the addition of haze made the FS-world even more realistic. More that 350 airports were dropped into it and refuelling made a hop across the pond much more feasible by utilizing the now standard refuelling stations all over the virtual world. Finally, a per- formance switch was added that gave the user a playoff between higher frame rates or better looks.

Also in 1995 BAO's long anticipated FlightShop arrived on the scene, but to some it came too late. The new co-ordinate system of Flight Sim 5.0 rendered all previous scenery and aircraft, created with A&SD, obsolete and something had to be done during the interim period between the release of FS5 and FlightShop. Fortunately, flight simmers from all over the world managed to understand (crack) the BGL scenery code of FS 5.00 and set about creating scenery generation programs for themselves. This lead to a stream of scenery becoming available on the Internet for download and file repositories began to build their databases. However, aircraft still proved to be a challenging prospect for designers and Flight- Shop was going to be the only program to ease their frustrations.

When FlightShop was released it gave designers the scope to create aircraft far superior to those found in Flight Simulator. In addition, FlightShop contained a few ready made aircraft, including a Boeing 747, DC-3, Beechcraft Baron and an ultralight. As more and more aircraft became available and designing techniques improved through "friendly competition" a flood of new planes appeared on the market. Some were even better than the default Flight Simulator aircraft themselves! The only drawback was the inability to create moving parts and default FS panels could only be used in conjunction with their super designs. Another benefit of FlightShop included modules for the creation of user defined adventures including ATC (Air Traffic Control) and flight plans. In essence Flight Simulator, for the first time, was becoming interactive. It really was a giant leap forward for the flight sim industry and was a major contributing factor to the add-on market we know and love today.

BAO taken over by Microsoft !

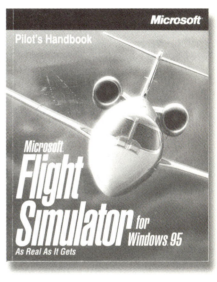

In January 1996, not long after the release of FS 5.1, Bruce Artwick sold both BAO and his copyright to Microsoft. As he pointed out in a column in Microwings Magazine, he was convinced that a small firm like BAO would not be able to generate the resources needed to survive in the ever demanding world of computer entertainment in general, and Flight Simulation especially. Most of the developers of BAO joined Microsoft. Bruce Artwick himself did not make the switch, but remained involved in the development of MS-FS as a consultant. Around the same time his former company subLOGIC was taken over by Sierra, another large publisher of gaming titles, to develop an up-and-coming and rival flight simulation title called **ProPilot** which later became a very popular alternative to Microsoft's Flight Simulator.

Generation 6

The very last version of Microsoft's Flight Simulator, developed by BAO, was Flight Simulator for Windows 95 (FSW95 or FS 6.0). This was released in 1996. According to Artwick, being an even-numbered generation, this would have been a refinement-release. In fact it was regarded as such as most of its improvements related to aircraft, fully-textured scenery and buildings. New aircraft were added: the EXTRA 300S together with instruction by aerobatic champion Patty Wagstaff, a Boeing 737-400 and Learjet 35A. Both included their own digital panel with new instruments and enhanced flight dynamics throughout.

This newer version demanded a bigger machine. The minimum being a 100MHz 486 or 60MHz Pentium with 8-16 Mb memory, 40 Mb of disk space and SVGA card with a resolution of 640x480 and 256 colours. Menus were adapted to a Window style. The most interesting thing however was that with the porting to Windows, frame rates improved by 20-50%. This was even with an upgraded resolution, contrary to what was expected by "Anti-Windows users", who recalled the problems of trying to run subsequent versions under Windows 3.11. In a column Bruce Artwick suggested that the passing of the 640K memory barrier and much faster graphic "blits" were THE major contributing factor to the overall improvement in performance under the Windows Operating System .

The next version, FS98 (internally called version 6.1), was introduced in August 1997 as the "15th year Anniversary" release. It boasted more than 10 million copies sold world-wide. This was seen as being a maintenance release. Nevertheless it included many new and enhanced features. The most important being a true rotary wing helicopter by way of the Bell 206B JetRanger III. This version brought a lot of handling improvements, compared to its predecessors and a much higher screen resolution too. It could now display up to 1280x1024 in 16bit colour.

Besides the helicopter there were a number other new aircraft. A Boeing 737-400 for one, a Learjet 45S with a brand-new glass cockpit and a glistening photo-realistic panel for the Cessna. An "extremely cool" addition was a Virtual Cockpit View which added depth but was short of working instruments. New sampled sounds added to the perception of reality. The navigation database was expanded to 3000 airports and a full FS installation needed at least 315 Mb of hard disk space.

Finally, a new option was added to give pilots the opportunity to "fly-online via the Internet" by way of a innovative and upgraded "Multiplayer" system. Utilising a TCP/IP internet connection a pilot could now choose between being a Player or an Observer. This system was snapped up by all kinds of groups. Real-time ATC FLY-ins and other online events began cropping up via Virtual Airlines many organisations were making the most of the novel and new fangled invention for other purposes. All kinds of auxiliary Network add-ons soon became available as well including Squawk Box and Pro Controller - to name a few. Microsoft joined in too by opening a special Flight Simulator "room", or "Gaming Zone" at http://www.zone.com.

Microsoft Flight Simulator 98

By 1999 Microsoft made it into the records books when 21 million copies of Flight Simulator sold worldwide!

Generation 7

In September 1999 **FS2000** was released in two editions: Standard and Professional. FS20002 marked another example of ground breaking programming as it featured a completely new 3D scenery engine. This time its was based on an elevation grid database. New and improved high resolution features added to the "realism factor. Coastlines, landmarks and seasonal textures were in abundance. This rendered all previous scenery add-on more or less obsolete, but improved the realistic look of the scenery by a factor of 150%. It was reported that more than 130 developers were involved in this version.

Other highlights were new aircraft including the King Air 350, Mooney Bravo, Boeing 737-300, Concorde (approved by British Airways) and a much improved Bell JetRanger III. All aircraft came with their own photo realistic panels and even interior "Virtual Cockpits" that could be viewed in "stepped" 360 Degrees.

Panels now included a GPS system as standard. Weather data could be downloaded from Jeppesen via the internet. Numerous navigation enhancements vastly improved the flying experience. An all new navaid database (also supplied by Jeppesen) now gave pilots a selection of over 21.000 airports to choose from. However, some were seriously corrupted and could not be used at all because of elevation data errors. Luckily, these were to be resolved in the next version.

Interestingly, many of the features incorporated into this version had originally been created as add-on products (freeware and payware) for Flight Simulator for Windows 95 and 98.

Unfortunately some of the biggest improvements in FS2000 proved to be its biggest weakness. Soon after its introduction customers began complaining about serious side-effects caused by the all new scenery engine. Vanishing roads, uphill rivers, fading runways and sinking aircraft were just some of the problems reported. Pilots later discovered that these symptoms were caused by the inclusion of the new navaid and scenery databases mixed with older, outmoded bits of programming. Aircraft and building shadows, landing light effects, among other things, were removed as well to the dissatisfaction of many. Removing features such as these were an utter crime to users but Microsoft claimed that it was a trade off in favour of better performance. A patch improved a few of the anomalies, but didn't keep everyone happy. So in March 2000 Microsoft published a second update that solved most of the problems. They even went further by making sure that shadows worked again which pleased many people. But alas, still no landing lights!

Generation 8

In October 2001 Microsoft's latest title, Flight Simulator 2002, was released. This again is a refinement generation. Nothing spectacularly new, but the list of improvements and additions are long and very impressive. Microsoft seemed to have listened to its customers and monitored the add-on industry closely. They now boast that this newer version is "created by real pilots".

There are new aircraft models including the Boeing 747-400, Beech Baron 58 and Cessna Caravan Amphibian and derivatives thereof. Each have stunning attention to detail. Many new special effects adorn these flyers too for example: virtual 3D-cockpit views with working instruments which can be viewed in conjunction with 3D Glasses. There is an independent and interactive ATC (Air Traffic Control) feature. New lessons and adventures are plentiful and really show off the new CFS (Combat Flight Simulator) based scenery engine. Even landing lights and shadows are back in abundance!

The most striking change is the addition of **auto-gen**erated scenery. As the aircraft fly above cities, farmland or other landscapes, Flight Simulator automatically renders buildings and vegetation appropriate to the scenery below. These objects appear smoothly from the horizon and fill the lush landscape with the type of detail that would have been impossible in previous versions. While **Auto-Gen** fills in ground details a new AI (Artificial Intelligence) system spawns air traffic between and around airports. The pilot is no longer alone and he can even see and hear other aircraft talking with ATC (Air Traffic Control) as he flies. If that isn't enough the pilot can even interact with ATC by way of key presses.

As a bonus, Microsoft seems to have been able to speed up and optimise internal calculations considerably. Never before has the performance of the aircraft and scenery been so fluid.

Undoubtedly, as with all previous versions, there will be improvements to make. This will be dictated by "virtual pilots" and the add-ons created in the months and years to come. Microsoft have already hinted at an opening for these by way "**gMax**" - a modelling tool for the construction of scenery objects together with an improved Flight Dynamics Editor for aircraft. So whatever can be next? The mind boggles!

The development of Flight Simulator has generated a lot of enthusiasm in the software industry over the years. It consistently remains in the top 40 of all software product sales. What was a "cottage" industry has matured vastly and is now worth millions (possibly billions) of dollars. As you will discover in the following pages there are countless individuals and dozens of companies who strive to make flight simulation "As Real As It Gets". However, all this would not have been possible without a man who's ingenuity and creativeness was THE catalyst for the development for one of the most popular consumer lead flight simulation products on the planet. A high-flyer named Bruce Artwick. We are forever grateful.

Bruce Artwick

Hardware
for
Flight Simulation

Hardware for Flight Simulation

The Great Hardware Dilemma...

So, you've purchased the latest and greatest flight simulation software, installed it onto what you thought was a powerhouse PC, and now you sit staring at what might be considered as a slide show. Right? You find that you have to turn down all the graphics and performance sliders in the sim to make the thing work and even then things don't seem to go the way you want them too. Well, before giving up on this great hobby and turning to knitting, lets see what we can do to improve things.

Most computer pilots find that they are in need or want of an upgrade at some point or another, whether it be to maximize their experience or to simply enjoy a smooth ride on a scenic flight along the coast.

Changing hardware components is a bit frightening for some. Indeed, we remember the first motherboard we replaced. An ABIT LX6 that allowed a bus speed boost from 66 to 75, 83 or 100MHz. It was an all night affair but once we talked to friends, read a review and drank lots of scotch we soon understood the basics and it really wasn't difficult at all!

Most people find abbreviated words like CPU, RAM, CDROM and other weird expressions like "Bus Speed" very daunting and tend to put them off before they even start. There isn't anything to them in all honesty. They are simply abbreviations to make tech-heads sound good. As soon as you get to learn what these expressions mean, you will discover that they play a vital roll in flight simulation but more importantly your whole computing environment.

"Don't throw away your PC because of its poor performance. A simple upgrade like a graphics card, CPU or memory chip can, in some cases, improve your virtual flying experience by up to 92%"

Today, video cards, motherboards, and hard drives are swapped and upgraded with great frequency, thanks in part to the evolution of operating systems and the easy way components fit together. Microsoft, along with hardware manufacturers, have made it a task almost anyone can perform, allowing you the user to squeeze more power from your PC without sparks flying all over the place.

"32% of bad graphics rendering is a result of outdated drivers. Try loading newer versions and you will be surprised with the results"

The ultimate question so frequently asked is; "What component do I spend my money on first?" It is no secret that processors carry the biggest burden in flight simulation, as complex calculations are carried out at any given moment - by the hundreds of thousands. However, putting all your eggs into one basket, as you might have guessed, can and often does create headaches.

Taking a perpetual ride on the CPU replacement bus without considering other critical parts is asking for trouble. Computers have evolved incredibly in a relatively short time (5 years), but one thing has remained constant; *a system is only as fast and stable as its slowest and weakest parts*. Figuring out just what component to consider takes some knowledge and understanding, which hopefully you will gain in the pages that follow.

These are just the basics. We encourage everyone to take some responsibility, learning their computer's strengths and weaknesses, and move methodically down the road towards trouble free flight simulation. Read books, go online and visit internet hardware web sites. Visit flight sim sites and ask other simmers in newsgroups and forums plenty of questions before embarking on any project. KNOWLEDGE IS POWER remember so It makes sense to do your homework.

Lets start by taking a look inside a computer and see what bits are in there.

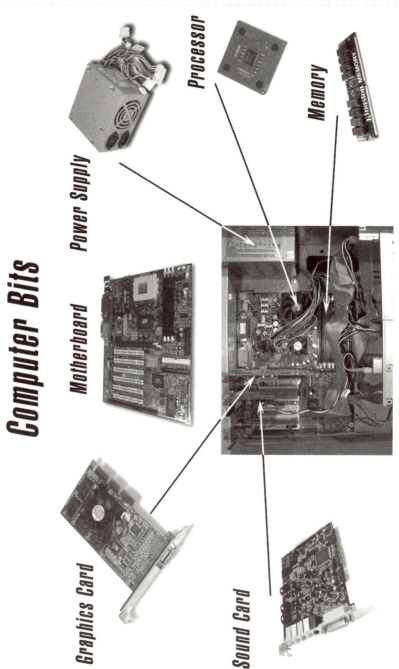

Computer Bits

Power Supply

Processor

Memory

Motherboard

Graphics Card

Sound Card

Motherboards...

The motherboard is the foundation of any system and ultimately determines overall system stability. The core logic chipset, CPU interface, expansion slots, and memory slots are but a few of the items that motherboards provide housing for.

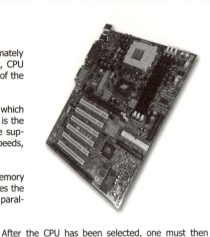

VIA, SiS, ALI, and AMD manufacture core logic chipsets, which control the flow of data throughout the board's circuitry. It is the design of these chipsets that determine which CPU will be supported, along with front side bus speeds, memory bus speeds, and AGP modes.

The "Northbridge" (VT82C694X) controls the CPU bus, memory bus, and AGP bus, while "Southbridge" (VT82C596B) handles the PCI bus, hard drives, power management, USB, and serial, parallel, and printer ports.

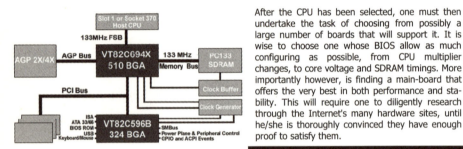

Diagram of the VIA Apollo Pro 133A chipset and the various connecting buses and components.

After the CPU has been selected, one must then undertake the task of choosing from possibly a large number of boards that will support it. It is wise to choose one whose BIOS allow as much configuring as possible, from CPU multiplier changes, to core voltage and SDRAM timings. More importantly however, is finding a main-board that offers the very best in both performance and stability. This will require one to diligently research through the Internet's many hardware sites, until he/she is thoroughly convinced they have enough proof to satisfy them.

PRIMARY CONSIDERATIONS
Adjustable BIOS. DDR CPU bus support Minimum 133MHz memory bus speed. Avoid on-board sound and video; slow system performance will prevail.

BIOS

Basic— Input— Output—System

Not limited to motherboards, BIOS are a set of programs encoded in read-only memory (ROM) chips handling such things as power-on self test (POST), setup configurations, and operating system service routines, such as interrupt requests (IRQs). Today BIOS are flashable, meaning manufacturers can update the BIOS software, perhaps fixing bugs, or enhancing

FACT FILE	
Motherboard **Main Board** **System Board**	**Chipsets** **Core Logic Chipsets**
The largest physical component sometimes thought of as the "heart" of a computer. Ultimately the foundation and most significant circuit board, around which all others are dependent.	A set of chips that collectively control (among other things) external buses, memory, and some peripherals. Located on the motherboard, its design determines what CPU will be supported

CMOS

Complimentary - Metal - Oxide - Semiconductor

A special battery powered chip, located on the motherboard that stores basic system settings such as the date, time, and system parameters. The BIOS reads the CMOS at start up, which then provides the setup routine the BIOS requires. The cell that powers the CMOS looks exactly like a digital watch battery. That's because it is a digital watch battery!

TIP: If your system time is faulty then look at changing your CMOS battery before doing anything else!

Power Supplies...

Often overlooked and taken for granted, the power supply's importance has escalated over the last 2 years due to the demands of new generation hardware.

Power supplies play important roles in the operation of PC's, converting AC (alternating current) from our walls into several levels of DC (direct current) which can then be used by electronic and electro-mechanical devices. As chipsets, CPUs, and video cards continue to evolve, the power requirements to operate them grows as well. Today, it is imperative when outfitting a computer to consider its power needs. Erratic system behaviours, such as intermittent system failures, are possible signs of a bad or inadequate power supply.

For absolute certainty, 300 watts is the recommended minimum for any new system or upgrade. Settling for less is simply skimping in an area that will bring nothing but trouble. Consider also the quality of the supply, normally reflected by its price. Bargains and great prices on can frequently be translated into inferior quality components.

Sound Cards...

Sound cards offer very little in the way of performance improvements, but are quite capable of adding incredible realism to the home flight deck. Not only can they rattle windows during reverse thrust but also provide a connection for a game controller.

In its basic form, a soundboard and its drivers translate digital sound signals (.wav files located on the hard drive) into analogue signals for playback and amplification through the speaker(s).

Many sound cards these days offer a variety of extra features for the discerning sound buff including Recording and Playback software, and enhanced gaming ports.

The main thing to watch out for is DirectX, BIOS and drivers. If not installed correctly a sound card can cause system crashes and can be your worst enemy. Double check operating system and driver compatibility.

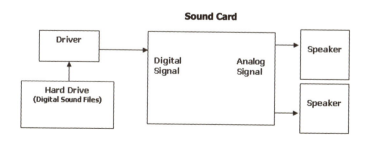

Sound Card

Driver		Speaker
Hard Drive (Digital Sound Files)	Digital Signal Analog Signal	Speaker

Processors (CPU)...

Unlike most 3D games, flight simulators are more dependent on the performance of a CPU than any other component. Of course we are referring to its rated MHz/GHz label that we all pay big money for.

Why is this so? Why do there exist so many games that are more graphics dependent, perhaps even memory bandwidth reliant?

Think for a moment about everything you are taking in whilst flying your favourite plane. The chances are the instrument panel has at least 12 functioning gauges, ever changing as you move along the airway. Furthermore, when you enter a 30-degree turn, things get even more interesting. Not only does a pilot need split second gauge inputs, but clouds, scenery, flight dynamics, G-forces, ILS or VOR distances etc. The list of variables is infinite.

Calculations are what it's all about, and for the smoothest experience from your system, this is worth repeating: *When we are talking flight simulators the processor and motherboard (therefore the chipset) ultimately determine the stability and speed of a system.*

There are several aspects to the architecture of a CPU that are important to understand.

Speed

Naturally how fast the processor runs is criteria NUMBER ONE!

Cache

There are usually 2 "levels" of cache the CPU utilizes and it is the amount and speed of this cache that we should be concerned with. Common amounts of Level 1 (internal) are 64kb to 128kb, normally functioning at the same speed as the CPU. Level 2 cache generally is 64kb to 528kb, with speeds that vary, but more commonly equal to CPU speed.

Front Side Bus

This refers to the speed that data flows from the chipset to the CPU and back. The fastest speeds today are 133MHz, 200MHz (DDR) and 266MHz (DDR). DDR (Double Data Rate) simply means that there are data transfers on both edges of each clock cycle, effectively doubling the data throughput. AMD first implemented this with the Athlon processor.

FACT FILE
Processor **Central Processing Unit** **(CPU)**
In the past referred to as the "Brain" of a computer, an integrated circuit consisting of an ALU (arithmetic logic unit), FPU (Floating Point Unit), and memory cache. Its duties include mathematical calculations, logical comparisons, and data manipulation.

PRIMARY CONSIDERATIONS
• Given the assortment of excellent processors and relatively low prices, here are some suggestions for your purchase.
• Minimum 1 GHz processor speed.
• On-board cache at least 256KB, full-speed, and 64kb of Level 1.
• Use DDR on its Front Side Bus, therefore at 200MHz minimum.

Memory (RAM)...

Memory constitutes the third level of performance in the flight simulation cosmos. RAM (random access memory) is the primary, yet temporary storage area for program code that the CPU will process. RAM is volatile; that is to say, it cannot retain information when power is removed, thus the need for hard drives (permanent storage).

This is yet another potential trouble area for gamers and power users; it is far too easy to be lured into purchasing bargain SDRAM modules, only to become ultimately frustrated when lock ups, Windows errors, etc. begin to rear their ugly heads. This says nothing of slower performance.

SDRAM

Synchronous **D**ynamic **R**andom **A**ccess **M**emory today comes in three basic speeds; 100MHz, 133MHz, and DDR (Double Data Rate). Memory bandwidth is a measure of the data throughput from the chipset to the memory address, affected primarily by this bus speed. As previously discussed, this is determined by the motherboard, its chipset, and the quality of the SDRAM module.

There is another facet of memory performance often neglected by the average consumer - Latency. Latency is the time from when an access command is issued (by the CPU) until the data becomes available. Latencies are a universal phenomenon and a main determining performance factor for the CPU cache, the HDD (access time), the PCI bus and any other component of any computer system, including the main memory. Current chipsets allow variations on two of these latency timing parameters:

- **CAS Delay (CAS): 2-3 cycles**
- **RAS-to-CAS Delay : 2-3 cycles**

CAS delay is much more important than RAS-to-CAS delay, and separates the men from the boys in terms of performance. Column Access Strobe latency which is set for 2 cycles can improve frame rates buy as much as 5-8% over a CAS setting of 3. Now, if you are not bored silly at this point, we get to the heart of the matter; High quality, more costly SDRAM modules generally are CAS 2 capable, whilst the generic bargain ones (CAS 3) stand no chance at this setting. Indeed, I have seen systems go absolutely black after adjusting CAS latency to 2 with an inferior stick of memory.

Knowing this, it is logical to assume we would attempt to build our system with the fastest bus speed support and lowest latency. Can all that memory bandwidth transform Fight Simulator xxxx? Indeed it will. Tests have shown 5-10% frame rate improvements using CAS 2 memory over CAS 3, in over 90% of the test scenarios. Only in non-intensive scenery (such as over open water or flat, remote areas of the globe) is the difference not noticeable.

Hard Drives (Disk Drives)...

Hard drives are permanent, secondary storage areas, consisting of rigid disks, coated with a magnetically sensitive material. These disks, manufactured with glass, are housed in a sealed case and together with read/write heads store and transfer a computer's data into RAM for usage.

A hard drive's performance is measured in access time, seek time, rotation speed, and data transfer rates. Twenty-first century drives are typically 20 to 30 gigabytes in storage capacity, with some manufacturers pushing well beyond the 100GB veil. Hard drives are commonly interfaced to the system through IDE (Integrated Drive Electronics) connectors on the motherboard, although SCSI (Small Computer System Interface) connections are sometimes utilized due to their extremely fast performance characteristics.

In simpler terms, this is the device that stores every single bit, byte, code, aircraft, program, what have you.

Is it essential to have a fast hard drive to enjoy a flight? No, a fast hard drive will load the simulation and any scenery, aircraft changes, etc. much quicker, but will not affect the actual frame rate during a flight. There are exceptions, such as a system very lean on memory (RAM) which might require Windows to access use its " virtual memory (Cache)". Virtual memory (Cache) is an area allocated on the hard drive to mimic system memory. The difference is one of access speed; system memory can be read and wrote to at tremendously faster rates than can a hard drive. Therefore, if virtual memory is required during a flight, some rather long pauses would be experienced.

Hard drives are remarkable feats of engineering; it has been said that the read/write heads are so close to the spinning disks, that it could be compared to a Boeing 747 flying 30ft above the ground at 500 Knots. This seems somewhat of an embellishment, but perhaps begins to aid in the appreciation of what goes on inside a hard drive case. The slightest imperfection or variation of the surface can spell catastrophe for the data contained therein. These may be the direct result of cigarette smoke contaminating the disk, or a sudden jolt to the case while the computer is operating.

Figure [2] shows the benefits of faster spinning drives.

Primary Considerations

Hard drive upgrades are done normally because of the need for more storage space, but it makes sense to consider a few things other than just bragging rights.

A high rotation speed is a sure-fire way to decrease the load time of all programs, including Windows itself. A speed of 7200rpm is notably faster than the lower cost 5400rpm, and is the recommended minimum (see chart above).

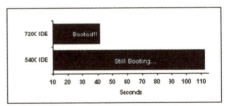

Figure 2

Data transfer rates also play a major role in drive performance. We are referring to the rate of flow between the drive and the controller (external data transfer), which currently is limited to 100mb/s. More commonly, Ultra-ATA drives may operate at 33mb/s or 66mb/sec all of which require a motherboard that supports these speeds.

Access time and seek time speeds are of secondary concern, as is the size of the buffer, or cache. Today most drives come equipped with 2MB of cache and access times of 8-9ms, which are quite satisfactory for the even the power-hungry user.

Graphics Cards (Video Boards)...

The source of disagreement and confusion, graphics cards (also known as boards, video adapters, VGA) attract more attention to would-be upgraders than perhaps any other hardware component. Partly due to marketing hype, partly to due misconceptions, there have been many disgruntled flight simulation enthusiasts after purchasing the video card purporting to " relieve the CPU's workload ".

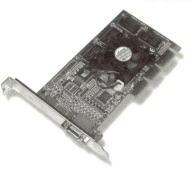

The reality is, no GPU (graphics processing unit) has yet given the overtaxed CPU a break that raises any eyebrows. That being said, the great improvements in visual quality are worthy of discussion and consideration. Full-Scene-Anti-Aliasing (FSAA) techniques are now standard features on today's graphics cards, and it is FSAA that transforms ugly, jagged 3D objects and scenery into smooth, realistic images on screen.

Primary Considerations

Today's accelerators will yield nearly the same performance in standard, non-FSAA modes, regardless of resolution. The challenge a prospective buyer faces is deciding on a product that grants excellent visual characteristics, reasonable price, compatibility, and maintains a certain degree of performance when all the "eye candy" features are enabled. This is quite a magical trick, if one can make it happen. Here are some basic features for suggestion;

1) Always purchase an AGP graphics board (card) .

2) Suggested minimum 32MB on-board memory; this will enable high resolutions in 3D, and allow for more video data to be stored close to the chip. With today's titles such as FS2002 we recommend at least 64Mb.

FACT FILE
Graphics Card **Video card** **Adapter** The circuit board that generates the output required to display text or graphics on a monitor. **Graphics Accelerator** A video adapter that includes a graphics co-processor, freeing the main CPU for other tasks.

3) Consider the 'balance' of visual quality and speed; one is useless without the other. Scrutinise the visual enhancements capabilities and whether they are useful for your particular preferences.

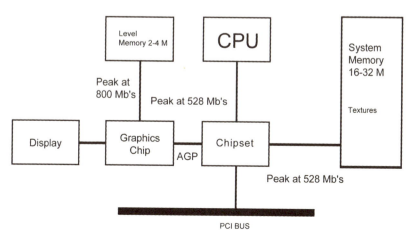

Monitors...

Of equal importance to the video card, today's high-resolution colour monitors provide us with sharp images, rich colours, and pleasing flicker-free displays.

CRT (Cathode Ray Tube) Monitors

Monitors are great! For a price!

Typical CRT Monitor

Monitors operate very similarly to televisions. The principle is based upon the use of an electronic screen called a cathode ray tube (CRT), which is the major part of a monitor. The CRT basically is a large vacuum tube lined with a phosphorous material that glows when struck by electron streams. This material is arranged into an array of millions of tiny cells, commonly known as pixels (dots). Three electron guns (red, green, and blue in colour monitors) situated at the rear of the CRT produce a controlled stream of electrons, beginning at the top of the screen and very rapidly scanning from left to right. They then return to the left-most position one line below and scan again, repeating this until the entire screen is filled. The electron guns are controlled by the video data stream coming into the monitor from the video card, which varies the intensity of the beam at each position on the screen. This control of the intensity of the electron beam at each dot is what controls the colour and brightness of each pixel on the screen. All of this occurs in micro- seconds, with separate streams for each colour coming from the video card. This allows the different colours to have different intensities at each point on the screen. By varying the intensity of the red, green, and blue streams, the full rainbow of colours is made possible.

The surface of the CRT only glows for a small fraction of a second before beginning to fade. This means that the monitor must redraw the picture many times per second to avoid having the screen flicker as it begins to fade and then is renewed. This rapid redrawing is called "refreshing" the screen, what you know as the re-fresh rate.

> ### Primary Considerations
>
> Besides the screen size, maximum resolution and its corresponding refresh rate are most important to scrutinise when selecting a new monitor. Resolution is defined as the total number of pixels in width and height, normally seen as two sets of numbers, such as 800 X 600 or 1024 X 768. As defined above, refresh rate is very important for eye comfort. A flickering screen (low refresh rate) can have adverse effects on the user's eyes, thus accelerating eye fatigue. A respectable monitor can display a refresh rate of 85Hz at a resolution of at least 1152 X 864. The dot pitch is also a consideration, however pitches below .28mm are very difficult to discern, thus making these figures more marketing hype than anything. The graphics card must also be taken into consideration, for your fancy new 21" monitor may display 1900 X 1600 at 85Hz, but your outdated video card might be unable to support these impressive numbers. The reverse scenario is also a possibility, where the video card has higher display capabilities than the monitor.

LCD (Liquid Crystal Display) Monitors

While CRT (cathode ray tube) monitors are the bread and butter of the PC market, LCD's (liquid crystal display) might be compared to a fresh garden salad. Lighter in mass and weight, healthier and more delightful to the eye, the advantages of LCD's over CRTs are tremendous, with only one noticeable drawback: cost. A 15" LCD monitor costs around £375.00 while its CRT counterpart can be taken home for about £200.00.

NEC Flat Screen (LCD) Monitor

> ### LCD Pro's
> 1) Flicker free display
> 2) Free of reflections and glare
> 3) Absence of pin-cushioning.
> 4) Low power consumption
> 5) Space efficient
> 6) Zero radiation
> 7) Superb colour/picture quality
>
> ### LCD Con's
> 1) Very Expensive
> 2) Requires special graphics adapter
> 3) Affected by cold temperatures

Laptop computers, pagers, and cellular phones are just a few vehicles that LCD's are used with, where weight and space are at a premium. TFT (Thin Film Transistor) is a type of LCD now commonly found on desktop monitors. Also known as active matrix, it refers to an efficient method of switching the display on and off, both in terms of current usage and image quality.

An important difference between CRT monitors and LCD panels is that the former requires an analogue signal to produce a picture, while the latter requires a digital signal. This fact makes the setup of an LCD panel's controls critical to obtain the best possible display.

Multi-Monitors and Networking...

The usage of multiple monitors has grown enormously over the years, the result of booming after-market software development. Moving maps, GPS displays, and multiple window views are just a few examples of the added enjoyment using two or more monitors can bring.

Two methods exist for multi-screen viewing; First, and possibly most popular, are with the use of graphics cards capable of outputting two displays or by adding a second card to the system. The second method, networking, requires two or more separate computer systems. Let's take a glance at what each has to offer.

Video Cards for Twin Monitors

Duel Output

Matrox G450 Duel Head Graphics Card

Operating system support for multi-monitors began with Windows 98, so it is no surprise that obtaining a video card is of no consequence. Matrox (www.matrox.com), Nvidia (www.nvidia.com), and ATI (www.ati.com) all produce graphics chips that have these properties. For the flight simming enthusiast however, there is a severe penalty with this method. Given the fact that flight sims are CPU dependent, the addition of another screen is often too much for one processor to handle. This method would be fine for a moving map or GPS, but certainly not adding a view outside the aircraft. For those that wish a wider view of the world from their virtual aircraft, truly the best option is through networking.

Networking

Networking involves connecting two (or more) computers with a special adapter card, cable, and in some instances a hub. Through special software applications, networking can offer the flight simmer much more than multi-monitors, as each system is capable of processing its own data, thus allowing for good performance on each screen. Primarily this technique is used to provide simultaneous views from the pilot's seat, making the experience that much more realistic and fun. Naturally, this is the more costly road to take.

Typical Network Card

Modems...

A contraction for modulator/demodulator, modems are the link between our homes and the world wide web (the Internet). A modem converts digital information into analogue sounds, which then travel over telephone lines, then translates that data back into digital form for computer usage. The great majority of PCs today are equipped with analogue internal modems, although newer, faster technologies are slowly taking hold. DSL (digital subscriber line) and cable modems offer much greater bandwidth than conventional modems, and would certainly be recommended for anyone interested in the multi-player features of many flight simulators.

3COM 56K LAN Modem

Common 56K modems connect to the PC in one of several ways, depending on the type. Internal modems plug into a slot (bus slot) in the motherboard, either ISA or PCI. External modems connect through a serial or USB port, which is also connected to the motherboard. Some people prefer external modems for a few reasons: they are easier to install into an existing system, they are equipped with status indicator lights, and you can reset them without having to reset the entire PC. Furthermore, an external is always a full hardware modem, rather than a 'host-based' modem. Host-based modems, also known as Winmodems (right), save money by replacing some of the processing capabilities with software drivers. In other words, it calls on the host CPU to perform the work, which can significantly affect overall performance. The drawbacks of external modems include increased cost and space requirements. Performance wise, the decision is an easy one. Hardware modems include

Typical Host-Based Modem (WinModem)

all processing capabilities in hardware on the modem itself, so for optimum performance, this is the best choice. If one doesn't demand too much from the system when the modem is operating, and the CPU is fast enough to handle the load, a software modem may be an acceptable way to economise.

DSL

DSL (digital subscriber line) is a high-speed connection that also uses the normal telephone line as its pathway.

DSL exploits the extra capacity that phone lines have to move data through the wire without disturbing the line's ability to carry conversations. With DSL, this line distance is 18000ft at maximum, and by way of its digital properties, offers increased bandwidth over a typical 56K. Most homes and small business users are connected to an asymmetric DSL (ADSL) line. Asymmetric defines the connection speed between the Internet and home. Since most users will download more than they will send, or upload, the connection from the Internet to the user is three to four times faster than the connection from the user to the Internet. DSL modems can mate to PCs through USB ports or through an Ethernet connection.

Cable Modems

The widely popular CATV (cable TV) network, like telephone lines, have much more capacity than is being tapped by the TV channels they were originally intended for. Direct cable connections to the Internet utilising a home's TV cable box offer arguably the sweetest solution of all. Cable Modems generally are much faster than DSL, with speeds from 128Kbps to 500Kbps and up. Access speeds of 1.5Mbps are possible, if local traffic is light. This perhaps is cable's only drawback, that users are sharing their bandwidth with each other. There may come a time when the roadway can no longer handle the traffic, thus slowing down data flow.

DSL PROS

• The Internet connection can remain open while still using the phone line for voice calls.

• Speed is much higher than a regular modem.

• DSL can use the existing phone line.

• The modem may be included in the installation fee.

DSL CONS

• The distance from the provider's central office affects DSL connections speeds.

• The connection is faster for receiving data than for sending data.

• Service is not available everywhere.

CABLE PROS

• Highest connection speed
• No dial-up required
• Full-time connection to the Internet

CABLE CONS

• Affected by local cable traffic

Controllers...

Joysticks, yokes, rudder pedals, and more recently 'switch boxes' might be held most responsible for bringing our flight simulation experience to life, more so even than the best graphics display or fastest CPU. These relatively simple devices allow a user to bring an element of hand-eye coordination to the virtual flight deck that would be somewhat impossible with only a keyboard and mouse.

Controllers interface with the PC through either via a Game or USB ports.

Joysticks and Yokes, like this from CH Products (right), are assembled with potentiometers, arranged perpendicular to one another representing two separate axes. Potentiometers are electronic devices that vary the resistance within a circuit, thus changing the amount of current flow.

The X/Y axes represent the roll and pitch of the aircraft, respectively. Today, several more axes are available, usually assigned to the throttle and rudder, with prop pitch and mixture axes becoming available on the latest USB controllers.

CH Products USB Yoke

Yokes, haven't been equipped with Force Feedback capabilities to-date. It is mostly Joysticks that have been given this functionality, such as Microsoft's SideWinder Force Feedback Pro (2).

Switch boxes like those made by Aerosoft (below) and GoFlight take realism to new heights by adding a third dimension of functionality by making systems like Autopilot, Trim, Landing Gear, NAV and COMM radio externally controllable with physical switching and visual monitoring via liquid crystal or LED displays.

Primary Considerations

Flight yokes certainly add a greater degree of realism to flying, unless of course the pilot prefers combat flying. Selecting a controller is more a matter of budget and user preferences. Professional style flight yokes can cost up to £1,000.00, but offer an exceptional degree of realism to virtual flight.

For several hundred less however, one is able to obtain an excellent arrangement of both yoke and pedals, and while not having the look and feel of a professional simulator, does an admirable job.

Aerosoft's ACP Compact

Joysticks come in many shapes, and sizes, some equipped with Force Feedback* capabilities, which add the dimension of vibration through the controller into the users hands.

Switch boxes can also be extremely expensive but do greatly assist the pilot when making rapid reaction inputs to the simulation.

Make sure you have done your homework when buying controllers as you can pay the same price of PC just for what most people think of as "bits and pieces".

What's Force Feedback?

Force Feedback appeared in the gaming world in 1997, bringing to the user the first physical sensations of bumps and jolts. These effects are the result of the controller simply moving itself.

How is this done?

The instructions that make a joystick move are created in software as waveforms (sound files), varying degrees of force over a period of time. The game software sends a token (a small burst of data) to the stick's onboard processor, which then can identify the waveform to use.

The processor then locates the wave, permanently stored on a ROM (read only memory) chip within the controller's base. The processor now has only to follow the instructions of the token, and send signals to two motors; one for the X-axis and one for the Y-axis. The motors are activated accordingly, pushing against the stick's axles and thus creating the force.

Hardware for Flight Simulation

Getting the most out of your computer...

Everyone seems to have answers for optimising their computer system. While they may be perfectly valid, more often than not the advice given usually relates to a particular hardware configuration. What follows is a simple 'common sense' checklist that can be used by many people and for a wide range of platforms.

Research

Total system optimisation begins with not only with quality hardware components but compatible components as well. How does one know? Hardware web sites exist for several reasons: To report when new hardware arrives on the scene and to then test them. Many questions are answered within reviews but more often than not they aren't read thoroughly enough.

It is very easy to rush ahead and post questions on newsgroups and forums rather than do a little research. Try visiting at least 3 web sites that have tested the device you are curious about. Then, perhaps visit flight sim related forums and search for posts that may be related to the product (s) you are interested in. You will be surprised how much more knowledgeable you become.

KNOWLEDGE IS PC POWER

Manual Scrutinising

Many hours of trouble and frustration can be avoided by calmly schooling oneself with the literature that is most frequently close to hand.

With particular regards to motherboards it is essential to understand, as much as one is able, the basic settings of a device. A motherboard's BIOS parameters affect stability and performance of the system it is installed upon. It is rare today to find one that has an ill-equipped manual. Read it, study it, know it, and feel free to experiment with different settings. Always record on paper what you have changed. The beauty of this is the fact that you can always revert to the original CMOS settings in the rare case a system fails to reboot. With regards to video cards, manuals are carefully written to define their driver installation and provide FAQ's for the device. Many hours of trouble and frustration can be avoided by calmly schooling oneself with the literature that is most frequently close to hand. Therefore, read it!

Maintenance

Scandisk, Defrag, Scandisk, Defrag: It cannot be said enough; disk errors frequently occur, especially if one downloads and installs new files on a regular basis. Running these tasks at least every two weeks is an essential and remarkably simple maintenance procedure that must be undertaken.

Turn off background applications like Virus protection and screen savers when using Flight Simulator. They only use up valuable memory and processing power when you need it the most!

Beware of Software!

This is a most distressing subject, but one that can be managed by being vigilant. Many software applications, upon being installed, audaciously plaster icons, help files, quick links and such all over your desktop - or even worse. Developers want your PC to show their logo and will do anything to achieve this. Some programs default to launching along side Windows wile others popup "nag screens" every couple of days. Not only are they annoying they slow down the boot process, rudely take up valuable memory space and occupy CPU cycles that could be constructively used for Flight Simulation. This is not good! Certain Anti-Virus programs and multimedia applications are prime examples of this. These should be configured to **a)** load at your command or **b)** close before launching any 3D programs. The purest form of Windows consists of "*Explorer*" and "*Systray*" which show in Task Manager after tapping the Ctrl-Alt-Del keys. You shouldn't see anything else. Note: This may vary if you are using an NT based operating system.

Spring Cleaning

Did you know that dust particles conduct current? Have you ever removed your PC's cover and gazed (or gasped) at the layer of dirt and dust on the printed boards that reside within? It is quite startling sometimes, especially if over 2 years has passed. Not to fear though - a good old-fashioned cleaning is all that is needed. Purchase 2 items:

1) A can of compressed air.
2) Static-free vacuum cleaner for PCs.

Both of these items are readily available at most hardware shops. First, using the compressed air, stir the dust up off the components, then vacuum carefully around the printed circuit boards, chips, fans, heat sinks, etc. until you are satisfied with the job. Performing this every 6 to 8 months will assure you will avoid any possible headaches due to lack of cleaning.

Cable Confusion

Wait - don't install that side panel yet! Take a look at your IDE cables and power supply wires. Are they neatly bundled, tied off? Are they impeding airflow? The chances are it resembles a bad dream, where one is entangled in a wild jungle-like entrapment of vines and branches. Yes, you say? Easily remedied by using plastic wire ties. Again readily available from your PC parts supplier. Carefully and methodically separate the wires and tie them into small bundles. People often attach a tie to the case. Then draw the bundles towards the case and away from the motherboard, chipset, and CPU. This vacates the air space in the centre, allowing for improved airflow and thus better cooling.

Take the ball out of your mouse every two weeks and clean it! Over 15% of computer crashes are due to dust particles clogging up the sensors that make it operate!

Conclusion

The final tip is not for silicon and metal, but rather for body, mind, and spirit, and a system greatly more complex and important than any computer shall ever be.

We speak of perspective, and while we abhor a lecture, we write this from the heart. We are dealing here with a COMPUTER! A device we believe is necessary to sustain life and flight simulation. A machine, that at times, we allow to control our emotions. Don't let it control you! You are the master (or should we say pilot) and it is you who should dictate how you want it to perform. Most of all, have FUN learning all you can about your computer and <u>don't</u> be afraid to experiment.

THE INTERNET & FLIGHT SIM

Flight Sim
and the
Internet

Flight Simulation and the Internet

If you are serious about getting into flight simulation then you are going to have to face the fact that at some point you will need to consider using The Internet. That's because 99% of anything developed, tested or discussed in flight simulation has originated from the good old world wide web. Fact! Colleagues gossip about it. People in Internet cafés slurp coffee over it. Your kids bang on about it. Lovers make music on it! Grandchildren leave you behind with it. You know the day is coming and there's nothing you can do about it! If honest you are fearful and it even makes you sick to the stomach to even think about it! There may be thousands of reasons why to avoid the Internet but please don't despair because you can be safe in the knowledge that statistics show that 60% of all computer users feel exactly the same as you. Even in Japan & China, where they are top of the tree in mathematics and electronics, people hate the idea!

The first and only reason why most "Arm Chair Flyers" get involved with the Internet is because they have purchased an add-on product for Flight Simulator and discovered while reading a manual the only way they can fix or update that software is to download a "patch" from the Internet. What a complete bummer hey? Most feel this is a stupid and manipulative con trick and think that all PC products should be supplied "free from defects". A valid argument, but unfortunately obsolete nowadays.

The problem is that in this day-and-age there are quite literally millions of combinations of computers out there. New technology appears on the shelves every day and innovations in design change the way PCs work by the flick of a switch. Developing flawless software is impossible because they will not work the way they should on all machines. A sad fact - but brutally true. An expression that immediately springs to mind is; "In my day, PCs would have been made from Rocks, Wood, Cast Iron and held together with 15 inch bolts."

Lets cut the chase and start from scratch by explaining why the Internet WILL play a vital roll in your flight sim life.

ONE GOOD REASON TO USE THE INTERNET...

A very convenient way for manufacturers to provide vital information about their software or hardware is via the Internet. It is speedy, straightforward and inexpensive for everyone. Developers can now give up-to-date (real-time) information about upgrades, fixes and other "tips and tricks" through special online support pages. A multitude of other information is usually thrown in for good measure too - and all for free. As soon as a developer creates an update he simply uploads file to his publicly accessible Internet server, e-mails his customers who then in turn visit the site to download it. Simple!

Traditionally, it would have taken a week to send a customer vital information about an upgrade/fix and cost tens of thousands in floppy disks or CDs. It now takes 1 minute to inform customers and typically 30 minutes for them to download it. So as you can see its quick, cheap and productive. CCP!

...MORE GOOD REASONS

You need immediate technical assistance to solve a serious incompatibility problem with software on your PC. The shops are shut, its Sunday night and you have a project that needs to be completed by tomorrow morning and your computer freezes every time you start the program. Luckily you can use the internet. The Manufacturer does not provide telephone support - which is useless anyway because it's the weekend and there is no description of the problem in the manual. Panic stations! All it says in the manual is "go online" to use the technical support knowledge base. What do you do? You use the Internet!

A friend lives 20,000 miles away and has discovered an extremely rare painting of a Prototype Supermarine Spitfire at an auction. He needs you to verify a photograph before he bids on it on your behalf. If he sends you a letter it will take a week to arrive. You have been waiting 30 years to get your hands on this painting! Have you got an e-mail address for him to send you the image before the auction stats in 2 hours time? What do you do? You use the Internet!

The Good Flight Simmer's Guide

You have heard that there are a limited amount of tickets for an Air show in the UK. They are 1/2 price and the show has your favourite Spitfire doing a flyby for the very last time. The tickets can only be purchased on the Internet. What do you do? You use the Internet!

You have heard that TecPilot (www.tecpilot.com) has copies of Flight Simulator for £55.95, £20.00 less than the recommended retail price, but you can only order them online. What do you do?

Roy Chaffin, an excellent FS aircraft designer, has created an accurate rendition of the Douglas RD-4 Aircraft for Flight Simulator to download. It has superior details and sound effects. You cannot buy it in the shops and it is not available anywhere else. What do you do?

Sold to the person in the Flak Jacket! However, before you can do ANY of this you MUST have a connection to the Internet via an ISP (Internet Service Provider). There is no other way to do it!

GETTING YOURSELF ONLINE...

To get online you must have an account with an "Internet Service Provider" (I.S.P). I.S.P.s provide you with a gateway or "port" to the Internet via a phone line. Typical examples of these include AOL, Freeserve, BlueYonder, BT Internet or NTL.

Through a cleaver device called a Modem, (see Hardware for Flight Simulation) your PC talks to a remote I.S.P.'s internet server (computer) which then allows your PC access the web and send e-mail messages around the world. Some I.S.P.'s are totally free while others ask you to pay a nominal fee each month. I.S.P.'s are a huge subject and really beyond the scope of this book so for now we recommend that you go and find an I.S.P. who can provide you with the services you require. If you have a cable telephone company call their customer service department now for information about "Getting Online" They may even send you a free disk that installs all the software you will ever need to get going! Take a look at our quick tips on the right for info on to getting started.

...The World Wide What?

The Internet is a fascinating place where one can get real-time flight sim news, reviews, downloads and articles - FREE! You must have seen and heard the infamous expression "Visit our web site at WWW DOT...." on TV and radio at some point? When you hear this expression all they mean is that you need to enter an address into your Internet Browser. Microsoft Internet Explorer and Netscape Navigator are the two most commonly used "Browsers" available today. Some Internet Service Providers supply their own software.

If you type WWW.**ANYNAME**.COM into your browsers address bar you will land on a web site. Try these examples when you first go online:

HTTP://WWW.AVSIM.COM
HTTP://WWW.FLIGHTSIM.COM
HTTP://WWW.MICROWINGS.COM
HTTP://WWW.SIMFLIGHT.COM
HTTP://WWW.TECPILOT.COM
HTTP://WWW.THEMAG-FS-NEWS-COM

Quick I.S.P. Tips

Get an I.S.P that supports 56k connection speed. Anything less will not do!

Find a modem that is not only capable of handling the internet but able to send and receive faxes. These are called "Fax Modems" They usually come with great software too!

Get an internet service provider that supports e-mail accounts greater than 5Mb. Anything less will mean that mail your friends send you gets rejected if you go over quota! Also make sure that their accounts DON'T have banners and adverts attached to your messages. Very annoying for your friends to have to be subjected too!

Don't use a voice phone line for your internet connection. Get a second line installed. They are cheap plus and you don't want friends complaining that your phone is always engaged do you? Plus if you want to upgrade your line to a hyper fast connection it will make the process much simpler.

If you have to pay for an Internet Connection to your I.S.P, and you have a separate phone line company, make sure the phone company knows that you are using the second line for the internet and fax calls. They may give you a substantial discount on call charges! Up to 40% in some cases.

Put a Dial-up Connection shortcut in the task bar. When you finish using the internet simply right click on it and select "disconnect". You'll be surprised how much money this saves you! It is very easy to forget to "Go Offline" and run up HUGE phone bills! Alternatively there are software timers available that will disconnect the connection if it is not used for more than a few minutes!

Get an I.S.P' that supports HTML formatted messages. If not you won't be able to send and receive colour messages with images embedded in them. Only plain, boring old and outdated ASCII text. Yawn!

Make sure your I.S.P gives you a minimum of 5 mail account aliases. Your wife and kids may need some for themselves and you don't want to die needlessly - do you?

Flight Simulation and the Internet

These are some of the biggest and most popular web site for flight simulation on the Internet. They are all great places to start with when getting into Flight Sim and several provide free add-ons to install into your flight simulator too! More about these in a moment.

In fact you should really type "HTTP://" before WWW. because some browsers will not allow you to type www. on its own. "HTTP" stands for Hypertext Transfer Protocol. The prefix WWW. evidently stands for World Wide Web, while .COM loosely refers to the type of institution you are visiting. Combined these quirky expressions are lovingly referred to as a Domain Names, URLs or Hyperlinks.

Most web sites make it easy for you to find your way around their pages by creating special links to click on with your mouse. **Some links can be bold, underlined sentences** while others can be images. Each of these "hot spots" are pre-programmed with an HTML (Hypertext Markup Language) code or a "hyperlink" to save you having to repetitively type links into your browser. Conveniently, more often than not, once you land on a web site you won't have to keep typing in addresses into your browser window again as most include links to other sites of a similar content.

E-mail (electronic mail) addresses work in a similar fashion to browsing links but instead of using HTTP:// WWW. These use real or nick names followed by an "at sign" @ plus the domain name of your I.S.P. suport@tecpilot.com is our mail address for example. To use e-mail you are required to use a different kind of software because it works using two different internet protocols called "POP" (Post Office Protocol 3) and "SMTP" (Simple Mail Transfer Protocol). Microsoft Outlook Express or Netscape Communicator are two of the most commonly used clients that support these formats. One is for sending the other is for receiving.

Browsing the Internet and sending mail are HUGE subjects and far beyond the scope of this book. We have provided a basic explanation but as you begin to use the Internet you will discover much, much more and will find it less daunting. Now we'll take a quick look at internet security and why that's important.

 ## INTERNET SECURITY...

The most talked about subject when approaching the Internet is security. Unavoidably, online banks and shops do get "hacked" and viruses frequently infect individual's computers. There is no getting away from it. Security is probably THE biggest issue on the Net and one must be cautious before embarking on any browsing activity.

When ordering a product online, and submitting your personal information, "HTTP://WWW." should be prefixed with HTTP**S**://www in your address bar. Notice the **S**? This means that the connection you are using is locked or secured - A SECURE connection.

1) You should always look to see if there is a yellow "padlock" present in your browser window.
2) Verify the sites "Certificate".

Your browser will provide details about Certificates this in its help file. Certificates are obtained by the administrator of a web site from an "Issuing Authority" - THAWTE is one of the most popular. A Certificate encrypts all the digital information being transferred between a user's PC and the web site server being accessed. This encryption usually takes the form of 128bit. Sounds complicated and you really don't need to know what that means but basically the higher the number the higher the encryption (security). Before embarking on ANY internet transaction you should ALWAYS check that the site you are buying from has a secure encrypted connection and that the connection has been checked and verified by a valid certificate issuing authority. Remember: **NO transaction on the internet is ever 100% safe!** Any web site that claims this cannot be trusted. In all fairness, Internet fraud is NOT common and the chances of you having your personal details compromised is around 20 million - 1 so don't let it worry you or put a damper on you ordering those essential air show tickets next summer!

Viruses, as they are affectionately called, are always in the news in one form or another. Trojans, Worms, call them what you like, these malicious little intrusions can bring entire networks to their knees. Not so much because of the damage they case but because of the propagation and subsequent mass e-mail warnings that follow. Viruses are routinely discovered on the Internet - rather like catching an infectious cold. Even as we write this section a major virus is infiltrating the web that sends a cascading file to all the mail addresses in an infected PC's contact list. This make people panic and immediately write to everyone they know to warn them about it. This then causes servers to crash due to the overwhelming volume of mail being distributed. Extremely inconvenient and most annoying to everyone on the net. You could call it Internet Terrorism.

The best way to steer clear of infections is to be conscious that they exist and to not be complacent. 60% of computer infections don't occur because of viruses infecting users machines automatically - but due to sheer user ignorance. Installing basic Virus protection software like that from Symantec (http://www.symantec.com) will lessen the chances of infection by up to 40%. Symantec is a market leader and must be one of the first places you should visit prior to using the internet. Norton Anti Virus is the most popular brand created for the PC by this company. But be warned, keeping your PC safe from Viruses and Hackers is down to YOU keeping responsible and alert. Maintain some kind of security on your machine ALL TIMES and examine your e-mails carefully for any unanticipated attachments before opening them. If you are suspicious DELETE THE SUSPECT EMAIL IMMEDIATELTY! Don't even think twice! Get rid of it fast and DO NOT use "mail preview" as viruses can be imbedded into a message and load automatically. When NOT using the internet (modem disconnected) and you have closed your internet and mail clients, switch your Virus protection **off** as it will reduce the performance of your machine and Flight Simulator quite considerably and lead to appalling performance.

SEARCH ENGINES...

As the name implies Search Engines help you to find things on the Internet. They are rather like using Yellow Pages and are really easy to use. Lycos (www.lycos.com), Google (www.google.com), Yahoo (www.yahoo.com), MSN (www.msn.com) and Hotbot (www.hotbot.com) are a few names to start with and each come packaged with many features. Some even provide free e-mail. Categorized into sections, they make the whole tedious process of finding things in the web very easy and

give you powerful tools to work with. However, some search engines have been criticized in the past for providing links to outdated or non-existent pages.

A typical advanced Search Engine system.

When using search engines, it's far better to use keywords rather than whole sentences. For example, if you are searching for a flight simulation magazine. Don't type; "I'm looking for the best flight simulation magazine on the planet" as this will result in a list of unhelpful pages containing every instance of the words: "I'm" -"looking"- "for" - "on" - "the" etc... This could amount to 80 million pages! Surely it would be much better to type "flight simulation magazine". This will reduce the search criteria by 50% and you are far more likely to find what you want. Some search engines even give you an option to add "wildcards" which allow you to narrow the search criteria by adding "AND/OR" expressions.

FLIGHT SIM WEB SITES...

We mentioned the top five flight simulation webs earlier but what do they actually do and how do they survive? As with any media these sites provide a daily fix of flight simulation specific news, reviews, articles and many other features -including free downloads. They do it just like any other news worthy source. They have members of staff who each fulfil a particular roll. Some are paid, while others are not. Some do it for the love of it as a hobby while others do it full time for a living.

CombatSim, MicroWINGS and TecPilot are what's commonly known as commercial based web site/magazines. They earn revenue from yearly subscriptions (internet and paper), advertising and shop sales. These organisations pay writers for material and because of this the public expects the content to be of a high standard. Which, on the whole it is. All content is accessible by inputting a special members password to gain access to protected "members" areas or "directories". By being totally commercial they offer special benefits to their members too. For example, discounts on products, days out, real flight simulator rides, free polo shits and other items. They are a fairly new concept on the internet and do offer a commercial perspective. Having said that, they fit nicely into a hobby by providing good value for money and they make every effort to work diplomatically with a community that dominates freely accessible content (freeware). In fact, recently MicroWINGS (commercial) and FlightSim.com (freeware) began to exchange news content on a regular basis.

Avsim, FlightSim, SimFlight and TheMag FS News offer "Freeware" services and can be accessed without charge. They tend to rely purely on advertising revenue and shop sales to survive. Staff who work for these sites are usually volunteers unless they are essential to an operation in which case they are compensated through free software or "pocket money". These large organisations operate using anywhere between 5 - 30 members of staff. Most tend to "employ" a team of regular writers who are mature, loyal and reliable and who generally provide well thought out material. Freeware content is continuously updated, news is provided on a daily basis (24 hours a day) and "How To" items are published in abundance. All provide some kind of public forum where users can express their opinions or ask questions and most have download repositories of files for various flight simulation titles.

Publications like TheMag FS News specialise in delivering solicited e-mailed bulletins. They "pool" news from top FS sites on a weekly basis. They tend to be balanced, co-operative, convenient and give enthusiasts a snapshot of the latest happenings around the world of flight simulation.

By pooling the news into recognised subscription based newsletters large web sites don't have to worry about "SPAMMING" people with duplicate or unwanted news. (SPAM: Unsolicited e-mail).

...SPECIALIST WEB SITES

There are many FS sites who primarily stick to a single activity. Some deal specifically with the development of add-ons while others build entire communities like "Virtual Airlines". Yes, there are even Virtual Airlines! Most provide downloads in one form or another and we suggest you flip to the section called Flight Sim Web Sits to see what's out there. But for now we'll take a look at the technical side of downloading files from the Internet.

INTERNET FILE DOWNLOADS...

There are many thousands of FREE downloads to be found on the Internet produced to a standard comparable to, if not better than some boxed products you will find in the shops today. To be fair to manufacturers some downloads aren't worth the time of day but occasionally you will stumble across a selection which are totally awesome and even better than commercially produced products!

Freeware "Authors" are people who make add-ons for themselves and then kindly upload them to the Internet for others to use for FREE!. Yes, this is a very popular part of the flight simulation genre and one that keeps the hobby alive and kicking. Before you go and download some wonderful files (and we lose you for a week) lets provide you with a few essential tips that will help you prior to downloading anything. We'll show you how to install them in the next section...

DOWNLOAD TIPS & TRICKS...

Firstly, pick a web site that checks its files prior making them available for YOU to download. This is very important because it has been known for people to provide files that are totally corrupt and full of viruses. Avsim.com, FSPlanet.com, SimFlight.com and FlightSim.com are a few respected sites that do this.

Next, check that you can see what the file size is before starting the download. Some sites provide the information as 26104kb or simply 26.1mb. But be careful of web sites that do not provide this information as you could spend hours downloading a single file. Here's a good tip:

For every 1000kb (1Mb) 15 minutes of download time (based on a 56k line).
For every 1000kb (1Mb) allow 7 minutes of download time (based on an 128K line)
For every 1000kb (1Mb) allow 2.5 minutes of download time (based on an 250k line)

If a README.TXT file comes with the download and is available to read **prior** to pulling down off the web ALWAYS READ IT. This simple document should provide a version history, installation instructions and give you a good knowledge of the designer or product. You can tell a lot from a README.TXT file. Check that it has the following items:

Does the author know what he is talking about?

> There is nothing worse than an author who creates a file and knows nothing about what he is talking about. Here's two extreme examples (true stories)
>
> **a)** A designer builds a Boeing 747-400 and suggests that flaps should be set to "half way".
> **b)** Someone creates London's Heathrow scenery. He suggests that runway 9R is 5000ft and has an ILS frequency of 112.5. Don't download it. It's the incorrect frequency and runway length.

Does the author give accreditation to other people's work?

> Some authors "borrow" certain files or "templates" to help in the construction of their add-on scenery and aircraft. If he has based the file on someone else's work then he should say so legally as it is copyright material.

Does the author have a web site you can visit and does the site work?

> There is nothing worse than finding out you cannot go to the author's web site to see if there is an upgrade, fix or work-around to a file that does not work!

Does the file include FULL installation instructions?

> If there aren't a basic set of instructions to install the product then there is really no point attempting to install it at all. Unless the file is a zip with a self generating directory structure. Well discuss that in more detail in a moment.

Is there contact information or an e-mail address?

> If you just cannot get the file to work OR you want to know more about other products the author creates then contact details are very helpful. If he doesn't provide them then he probably has something to hide OR is a hermit.

Is the download compatible with your operating system or flight simulator?

> This is probably the most important part. There is absolutely no point in downloading a file that ISN'T compatible with your version of flight simulator or operating system, is there? So make sure you check and double check that the file is compatible with whatever you intend using it on. If there is nothing mentioned then DON'T download the file.

Do you need to download other files to be able to use it?

One of the most annoying things to happen is when you install an aircraft or scenery is only to find your computer crashes when you use it. More often than not this is due to files that should have been installed prior to use. Make sure that the README includes ALL of this information. It is sometimes omitted because authors think we are all psychic. If you later write to the author following a problem and he says X Y and Z files should be installed then he needs a smack around the face with a wet fish! He should know better! Tell him you are NOT happy and he should update the file immediately before it happens to someone else!

If the person, who created the file, has not provided at least half of this information above, we recommend that you DO NOT download the file from the Internet. If the developer cannot be bothered to write a few simple instructions then he has taken shortcuts elsewhere. For example, he may of forgot to include landing gear, flaps or an essential file to make the aircraft work at all!

FILE TYPES...

One of the most common questions we get asked is, "How are downloads transmitted over the internet?".

Well, the most common way is for designers to bundle their files together so they download quickly is to create a file called a ZIP. A **ZIP** is a compressed folder that when executed "extracts" (un-compresses) the integrated files into directories chosen by the user OR the developer. Some zip files are specially formatted to extract automatically into the correct folders or directories. This is great for novices because it is simple a case of clicking one button and the rest is done automatically. You will need a program like Winzip (http://www.winzip.com) to be able to use these types of files.

Setup.exe

The other type of file is the **EXE**cution type. These are the simplest to use and we have all seen them in one form or another at some point. They are usually called **SETUP.EXE**. These types are usually employed when crucial changes are required and the programmer believes you will get wrong if you attempt to do it yourself. Windows registry settings are one example or additions to the Flight Sim configuration files. These setup routines act like just like ZIPS in that they have compressed files within them that are uncompressed (extracted) but they have an added ability to show the user how the setup is going through a progress bar and issue multiple commands, with help from the Windows operating system.

These SETUP.EXE routines are also employed when registration is required for "Shareware" (Try-before you buy) packages. They automate the process of filling in forms, going online or checking the length of time a product is due to expire. They are complex to use if you intend programming them and we don't suggest you get involved until you are totally competent with your computer. But of course, use them to install things by all means!

DOWNLOADING... (WAIT FOR IT !)

Now you have a basic understanding of how the internet works, lets see what sites are available and what files to look out for. Shall we go and download something now? No doubt you will have found yourself an I.S.P. so lets get cracking and find out what files are available for download! Start by visiting HTTP://WWW.AVSIM.COM and at the top of the page there is a link to their **Files Library**. Hey, while you are there you should check out their really great... Actually - we think its about time YOU started exploring for yourself - this is the easy part! See you in a while! (Also try some sites listed towards the back of the book)

When you have found a file (and remembered to read the README) click on the link where it says you can download it. When the windows download popup box appears choose, "SAVE". Press the browse button and locate somewhere on your hard drive that's easy to remember, for example: **C:\My Documents\Downloads.** Click on "OK" twice and the download will commence.

When your download has completely finished another pop-up dialogue box will appear that says, "Download Complete" . Click "OK" again and you can then navigate your way to the place you saved it.

Then, double click on the file to open it and locate the README.TXT file for further instructions. If you cannot open the file then you will probably need a copy of Winzip (currently Version 8.0 as we write this) from http://www.winzip.com to open/extract (decompress) the file you just downloaded.

If you need more advanced information about setting up your download in flight simulator then check out the following pages on installing Aircraft, Scenery and Panels - in the next section.

Also, check out some of the various web sites we mention above, plus the listings in Flight Sim Web Sites section later in the book.

The main thing to remembers is **don't be afraid of the Internet**. You are the one in control at all times. It can be great fun, a good learning experience and there are loads of super people who can help you get the most out of the Internet and Flight Simulation. Don't be afraid to ask plenty of questions in Newsgroups and Discussion Forums. The flight sim **"community"** both commercial and freeware are more than happy to help. After all, they were once where you are now!

Most importantly **HAVE LOTS OF FUN** experimenting and enjoy exploring the Wonderful **W**orld **W**ide **W**eb!

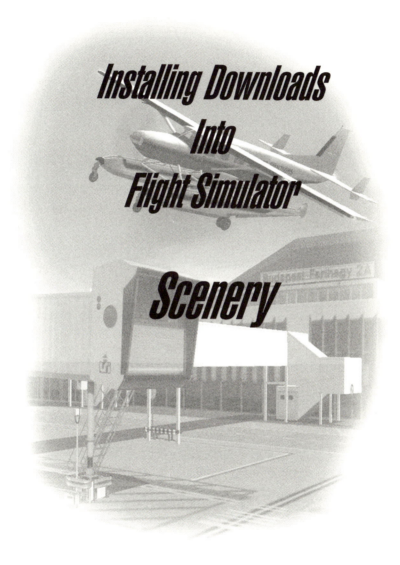

Installing Downloads Into Flight Simulator

Scenery

Installing Scenery into Flight Simulator

After flying around the simulated world of Flight Simulator for a while, many simmers decide they want to add some new scenery. Perhaps your favourite airport does not have the detail you would prefer or you would like a higher resolution for a particular area.

There are hundreds of add-on sceneries, ranging from commercial packages, to shareware and freeware. One of the first projects for many simmers is to add an airport. In this section, we are going to install new scenery for Dunkeswell Airport (EGTU), in Devon, UK by Paul Roberts and Mike Vernon. If you want to follow along on your computer, you can download it from www.FlightSim.com, a wonderful flight simulation site that has one of the largest files repositories in the world (note that you need to register free with FlightSim before you can download files).

We shall go though the process of installing the airport and along the way discuss some general information about scenery files. Conveniently, Flight Simulator has a good reputation as being backward compatible. For simplicity, this means that the installation processes is more or less uniform across different versions. FS98, FS2000 and FS2002 for example. In this installation we are using Flight Simulator 2000.

Once you have downloaded the file, unzip it in an empty temporary folder. Now let's take a look at the files and folders as seen in **[Figure 1]**. You will note that there are several files and a new folder in your temporary folder. As always, we shall start with the README.txt file, which usually contains a description of the scenery and some installation instructions.

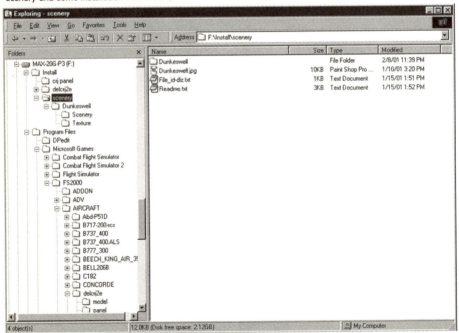

Take a look at the installation instructions included **[Figure 2]**. Read through them carefully. If this is your first time installing scenery, don't worry if things seem a little confusing at first. We'll discuss the details after you read the instructions.

Now, if we haven't scared you off, you may be asking yourself, "What did he say?" At the bottom of the instructions, you will see that we need two more files, Airport 2.xx and VOD3.0 textures. These are the texture

Figure 2

files supplied with two scenery design packages and are widely used by scenery designers. So that you are not forced to download them with each scenery package, the authors usually only tell you that you need them.

The two files you need can be found on this site by clicking the links below OR by visiting the files section at FlightSim.com .

aip210tx.zip Complete package of textures that accompany v2.10 of the Airport scenery design program by Pascal Meziat, Brian McWilliams and Tom Hiscox.

vodtex30.zip This file contains all genuine VOD textures for VOD v3.0 scenery. By Rafael Garcia Sanchez.

While we are at it, and if you plan to install more scenery, you might as well download the ASD textures also. Once you have these three installed, you will not have to download them again and they will be shared by any scenery that needs them.

asd21txt.zip These files are required for proper function of scenery designed with Abacus' "Airport & Scenery Designer" Version 2.1 (see Boxed Products - Utilities).

Download at least the first two and put them in another temporary folder. Next, unzip aip210tx.zip into the main Flight Simulator texture folder. If you installed using the default setting in Flight Simulator 2000, the path will be C:\Program Files\Microsoft Games\FS2000\texture. Next unzip vodtex30.zip into the temporary folder. You will now see vodtex30.exe. Double click on this file and enter or browse to the main Flight Simulator folder when prompted. You have now installed the support files for the scenery.

Let's now look again at the Dunkeswell scenery files we unzipped earlier. In the main folder, besides the Readme.txt file we have already looked at, you will find Dunkeswell.jpg, a screen shot of the airport and the conventional File_id-diz.txt, which is a brief description of the contents of the zip.

You will also see the Dunkeswell folder. Beneath the Dunkeswell folder, you will see a scenery folder and a texture folder. These two folders will be found in any scenery package and Flight Simulator will automatically look for the required files there. If you look in the scenery folder, you will see four files - DUNK_exc.bgl, Dunkeswell.bgl, Dunkeswell.xcl, and filelist.dat.

Let's digress a moment and talk about the flight simulator world. The flight sim Earth model is designed to be entirely covered with scenery. The scenery doesn't have to include a lot of detail everywhere, but wherever a user can travel there should be some sort of scenery.

Well defined and highly detailed areas are quite easily added using standard graphics language systems to create object scenery. You can use textured polygons, lines, and points to model an area to reflect the degree of detail that you choose. Flight Simulator uses a graphics language-based scenery system called the Object System, also known as the BGL Graphics Language.

The origin of the .bgl file extension goes back to the original Flight Simulator by Bruce Artwick. Some people claim that BGL actually stands for Bruce's Graphic Language, while others claim it is Binary Graphics Language. Whichever it is, the .bgl files contain the instructions used by Flight Simulator to draw the polygons, lines and points, as well as what textures to apply. The texture files themselves are bitmaps (.bmp) and the textures for the specific scenery area for Dunkeswell are found in the texture folder.

The two other files in the scenery folder are specific to Airport and VOD. In the texture folder, we shall find the .bmp files used to create the scenery.

Installing Scenery into Flight Simulator

We are now ready to move the scenery to Flight Simulator. There are two schools of thought on where to place the add-on scenery files. One group advocates putting everything in the directly in the Flight Simulator scenery folder. The other school, to which we subscribe, recommends putting add-on scenery in its own folders. The latter method makes troubleshooting and removal a lot easier. We also prefer to organise scenery in geographical areas. Remember, this holds true for later versions of Flight Simulator as well.

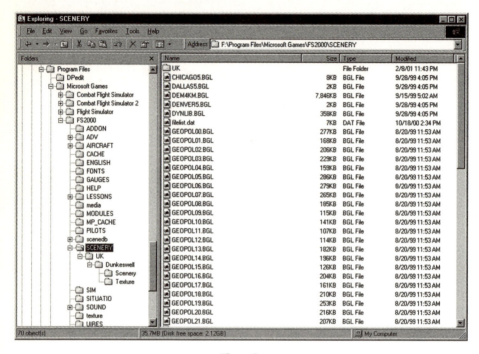

Figure 3

First create a UK folder under the main scenery folder in Flight Simulator.

Then, either copy or move the Dunkeswell folder and the two folders beneath it to the FS2000\scenery\UK folder. **[Figure 3]**

It is not necessary to move the readme.txt, Dunkeswell.jpg and file_id-diz, unless you want to keep them for future reference.

When you are done, you should see the Dunkeswell folder beneath the \FS2000\scenery\UK folder as well as the scenery and texture folders beneath the Dunkeswell folder.

The next thing we need to do is to tell Flight Simulator where to find the new scenery. To do this, we will add a Layer in the Scenery Library and also an entry in the scenery.cfg file showing Flight Simulator where the scenery is. More on that in a moment.

Before we actually install the scenery, let's take a look at the scenery.cfg file. If you don't know how to open this file then here's a quick lesson.

Go to you main Flight Simulator root Directory - this is usually found at:

c:\program files\microsoft games\fs2000\scenery.cfg

When you see the file it might not have the extension ".cfg" , just the word Scenery. In this case you will have to tell Windows to show the file extension. This can easily be done by:

Opening My Documents or Windows explorer and select: View > Folder Options > and select the View (tab)

In the text area, UN-CHECK the box that says "Hide file extensions for known file types".

While we are here lets make sure that the radio button that says "Show all files" it is set to **ON**. This is just in case you cannot see the file at all when you go to the FS2000 directory. Sometimes applications hide files to prevent users tampering with them. **[FIGURE 4]**. Now Click OK and return to the FS2000 directory where you should see the entire file name including its extension (.cfg)

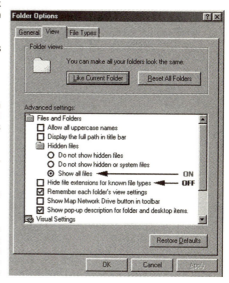

Figure 4

Figure 5

Now Double Click on your Scenery.cfg file and it should open in Windows NotePad. If not then a dialogue box will appear informing you that the file is not associated with anything. **[Figure 5]**

We know it really does associate with Flight Sim but Windows does know how not open it outside the simulation. So what we need to do is ask Windows to associate the file with Windows NotePad. Then we will be able to open it for editing. Therefore, scroll down the list of applications until you see NotePad - then select it. While we are at it, so we can edit the file later, also check the box that says, "Always use this program to open this file". Click OK and then the file will open.

Now then, let's take a good look at it...

The scenery.cfg performs several functions in Flight Simulator. It sets the maximum Cache size, whether to empty the Cache upon exit, and where scenery files are located. It can also be used to exclude certain types of scenery or flatten the terrain mesh in a given area to a certain elevation. The scenery.cfg replaces the world.vis files used in previous versions of Flight Simulator.

Installing Scenery into Flight Simulator

Scenery area entries tell Flight Simulator where scenery is located. A typical scenery area entry in the scenery.cfg looks like the following:

```
[Area.nnn]
Title=Somewhere Scenery
Local=C:\Scenery
Active=TRUE
Layer=nnn
```

The descriptions for the parameters contained within the scenery cfg file are described in detail on the next page.

Entry Title: `[Area.nnn ]`

Where nnn is the scenery area entry number. nnn must be an unique number.

Scenery Title: `Title=aaaaaaaa`

Where **aaaaaaaa** is the title of the area. This text will be shown under the "Scenery Area" column in the Flight Simulator 2000 Scenery Library.

Path Designation: `Local=path or Remote=path`

When using the "Local" designation the scenery is read directly from the path. When using the "Remote" designation the scenery is copied from the path to a folder created automatically by Flight Simulator in the "Cache" folder in the main Flight Simulator directory. The scenery files are then read from the new folder in the "Cache" folder. The "Remote" designation is useful when scenery files are in a location that cannot be accessed as quickly as the local hard drive such as CD-ROM drive or remote server.

Both "Local" and "Remote" Paths can use a variety of formats for designating the path to the scenery.

They are:

Drive:\path - the typical path designation (e.g., C:\MyScenery)

[Volume]\path - this method uses the volume name of the drive and then the path (e.g., [MySceneryCD] \MyScenery)

\\Machine\share\path - this is a server path (e.g. \\MyServer\MyScenery)

Status: `Active=TRUE or FALSE`

A "**TRUE**" setting tells Flight Simulator to display the scenery and a "**FALSE**" setting tells Flight Simulator not to display the scenery. This is shown in the Flight Simulator Scenery Library under the "Enabled" column.

Layer Number: `Layer=nnn`

Where nnn is a number typically the same as the Entry Title area number. Higher numbered layers have priority over lower numbered layers. This number is shown in the Flight Simulator Scenery Library under the "Priority" column.

Miscellaneous

`Texture_ID=0` and `Texture_ID=1` are special settings that should usually NEVER be changed. `Texture_ID=0` sets the global default texture directory to be searched if the requested texture cannot be found in its sister texture directory. This is also the location of Flight Simulator's global textures that have no sister texture directory. `Texture_ID=1` is the default terrain texture directory.

Exclusion and Flattening switches

Some backward-compatible/add-on sceneries may not appear correctly in Flight Simulator. There can be several reasons for this. The reasons most often responsible for such problems are:

Flight Simulator uses a digital elevation mesh terrain system instead of the flat "seed" system used in prior versions, and Flight Simulator incorporates nearly every airport and published navigational aid in the entire world.

Flight Simulator also has two "switches" - Flatten and Exclude - that can be used in the scenery.cfg file to solve many of the visual problems with backward-compatible add-on scenery. The Flatten switch flattens a four-sided area to a single elevation. The Exclude switch excludes the default scenery in a four-sided area from being displayed or transmitted in the case of radio navigational aids.

Both switches affect scenery areas with a lower layer number than the scenery area they are added in. For example, if the switch is placed in layer 50, then only layers 49 and below will be affected. Switches are added to the end of a scenery area entry in the scenery.cfg file using a text editor. Exclude and Flatten switches can both be added to the same Scenery area.

Flatten Switch

The Flatten switch syntax is as follows:

`Flatten.X=Elev,La1,Lon1,Lat2,Lon2,Lat3,Lon3,Lat4,Lon4`

X is equal to a number between 0 and 9, and must be used in order, starting with Flatten.0, then Flatten.1, and so on, up to Flatten.9. You are allowed a maximum of 10 Flatten switches per Scenery Area.

The Elevation is in feet above or below mean sea level (msl) and can be any number between -2000 and 99999.

The Latitude and Longitude of points 1, 2, 3, and 4 must be entered in a clockwise or counter clockwise fashion around the four-sided area and must be in the form of degrees and minutes. You do not need to add the * or ' symbols to indicate degrees and minutes. The shape does not have to be a rectangle. Any four-sided shape will work. The maximum size of a flattened area is 90 degrees of longitude and 45 degrees of latitude.

Exclude switch

The Exclude switch syntax is as follows:

Exclude=North Latitude,West Longitude, South Latitude, East Longitude, category

Exclusion happens only in a rectangular area and must follow the form North Latitude, West Longitude, South Latitude, East Longitude. The Latitude and Longitude entries must be in the form of degrees and minutes. You do not need to add the * or ' symbols to indicate degrees and minutes.

The category determines which default scenery type you choose to exclude in the defined area. There are 4 categories:

objects - Excludes all default 3-dimensional buildings and objects as well as airports.

`vors` - Excludes all default VOR and ILS navigational aids
`ndbs` - Excludes all default NDB navigational aids.
`all` - Excludes all default objects and navigational aids

You can use one or more categories in an Exclude switch and the maximum size of an Exclude area is 90 degrees of longitude and 45 degrees of latitude.

Installing Scenery into Flight Simulator

Getting your installed scenery working in Flight Simulator

There are two methods to install scenery in Flight Simulator. The first method is the most simple and will work with quite a few add-ons. If a flatten or exclude switch is needed, you will have to use the second method. The process of adding scenery is the same for both methods up to the point at which we are now with the Dunkeswell scenery.

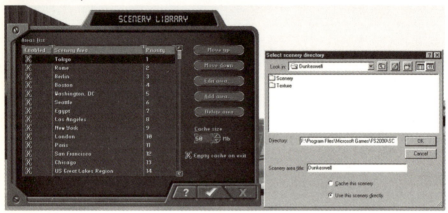

Figure 6

First Method

Move the scenery into \FS2000\scenery as we have done. Then start Flight Simulator and go to the 'World' menu. Click on 'Scenery library,' then click on 'Add area'. Scroll through the sub-folders until you see the one named 'Scenery' and double-click on it. Find the sub-folder with the name of your new scenery and double-click to open it. Click on the scenery folder inside the new scenery folder so that it is selected. Then type a title for scenery in the dialogue box named 'Scenery area title'. **[Figure 6]** Click on OK and back out to the cockpit view. There will be a pause while the scenery is added. When it is finished, you should be able to click 'World' and then 'Go to Airport' to select your new scenery. **[Figure 7]**

Figure 7

Second Method

This method is the one to use if either flatten or exclude switches are needed, as they are with Dunkeswell, and requires editing the scenery.cfg file. If you installed a new airport using the menu method described above and there was a default airport in the same location, you might see double or floating runways or other strange anomalies.

The initial steps for this method are the same as the first. After you have completed the steps to add the scenery in the first method, and before going to the new scenery, shut down Flight Simulator. We now need to edit the scenery.cfg file, which is found in the main Flight Simulator folder, and add the flatten switch. You can edit the scenery.cfg file with a text editor.

The authors of the Dunkeswell scenery have supplied the necessary entries for the scenery.cfg file in their instructions. We will add the flatten switch to the Dunkeswell section in scenery.cfg file.

NOTE: The Flatten switch line wraps on this page. Make sue that this and other lines do not have any line breaks in it when including them in the Scenery.cfg file.

```
[Area.xxx]
Title=Dunkeswell
Local=SCENERY\UK\Dunkeswell\Scenery
Active=TRUE
Layer=xxx
Flatten.0=850.9,N50 51.92,W003 14.64,N50 52.00,W003 13.58,N50 51.45,W003
13.62,N50 51.28,W003 14.58
```

Before editing the scenery.cfg file, it is always a good idea to save a copy of the original file in case you really mess up something and Flight Simulator crashes or won't start at all.

The last entry in the scenery.cfg file should be the Dunkeswell section and it should look like the following (the numbers may be different).

```
[Area.073]
Title=Dunkeswell
Local=SCENERY\UK\Dunkeswell\Scenery
Active=TRUE
Layer=73
```

Copy the entire Flatten.0= line from the installation instructions and paste it after the Layer= line. The entry should now look like the example below.

```
[Area.073]
Title=Dunkeswell
Local=SCENERY\UK\Dunkeswell\Scenery
Active=TRUE
Layer=73
Flatten.0=850.9,N50 51.92,W003 14.64,N50 52.00,W003 13.58,N50 51.45,W003
13.62,N50 51.28,W003 14.58
```

Save the scenery.cfg file and then close Notepad.

Now start Flight Simulator. Once it is up and running, go to the "World" menu, then "Go to Airport". Set the Global selection for Europe, the Region as United Kingdom and then enter Dunkeswell in the Airport dialogue box or scroll to Dunkeswell and click. **[Figure 8]** Click the green checkmark and you should be taken to the newly installed Dunkeswell Airport.

Taking a few minutes to download and install add-on scenery can make a big difference. It enhances the visual experience by adding more detail and realism.

There are many graphics applications that enable you to design scenery too. Take a look at Lago's Scenery Enhancer for FS2002 for example. We will cover building scenery in the next "*The Good Flight Simmer's Guide*".

Figure 8

Installing Downloads Into Flight Simulator

Aircraft

Installing Aircraft into Flight Simulator

For this exercise, we shall install a Canadair CRJ200-ER, which you can download from FlightSim.com. The filename is delcrj2e.zip. While you are at it, also download cl65panl.zip too, which is the panel we will install in the next chapter. Remember that even though this tutorial covers an installation for FS2000, it can be applied to FS2002 as well.

Though many downloads come with instructions such as "unzip the file into your main Flight Simulator folder", this can often lead to problems. We have seen zip files with incorrect (or no) paths that spattered files all over the place, creating a mess. Therefore, you should always unzip the file to a temporary folder. When you extract the files with WinZip, make sure the option to "Use folder names" is checked. This should create the proper folders and put the files where they belong.

When the extraction is completed, you will see a new folder beneath your temporary folder with the name delcrj2e. Beneath delcrj2e you will see eight files and four folders - model, panel, sound and texture. **[Figure 1]** One of the files is readme.txt. This file or one with a similar name is the first one you should take a look at when installing a new aircraft.

Double-clicking the file should open it in Notepad or some other text editor if you have replaced Notepad. In the readme.txt, you will usually find any installation instructions. For the CRJ, the instructions are brief:

To install: UnZip to your FS2000\AIRCRAFT Folder using WinZip or similar program making sure the "paths" option (Use Subfolders) is turned on.

Before we actually install the CRJ into Flight Simulator, let's take a look at the files, folders and their purposes. Knowing the purpose of the files and what they contain will help in troubleshooting any problems once we get the aircraft installed in Flight Simulator. Starting with the files in the delcrj2e folder, we find:

Figure 1

The Good Flight Simmer's Guide

The main aircraft folder

aircraft.cfg

This file is required for all aircraft. Open it in Notepad and take a look. (like you did with the Scenery.cfg file in the last chapter)

```
[fltsim.0]
title=DelawAIR Canadair RJ200-LR
sim=crj5kend
model=
panel=
sound=
texture=
checklists=
kb_checklists=CRJ2k_CHECK
kb_reference=CRJ2k_REF
```

[fltsim.n]
Represents a different configuration of the aircraft, and is known as a configuration set. If there is only one section, it will be labeled [fltsim.0]. If there are multiple configurations, they will be labeled [fltsim.0], [fltsim.1], [fltsim.2], and so on. Each one refers to a different version of the aircraft. Using multiple configuration sets allows a single aircraft container to represent several aircraft, and allows those aircraft to share components.

title= Determines the name that will appear in the Flight Simulator aircraft listing.

sim= This is the filename (without the extension) of the .air file used by the aircraft. If you have multiple ,air files for the same aircraft, each .air file would be listed in a separate [fltsim.n] section.

model= Identifies the model (.mdl) file. If there is only one, nothing has to appear after the equal sign. If there were multiple models, then each would be listed in a separate [fltsim.n] section.

panel= Identifies the panel file. You may have multiple panels by using multiple [fltsim.n] sections.

sound= Is the identifier for the sound files. You may have multiple sound sets by using multiple [fltsim.n] sections.

texture= This is for the texture files. These are the paint schemes, or liveries for the aircraft. You may have multiple liveries by using multiple [fltsim.n] sections.

```
checklists=
kb_checklists=CRJ2k_CHECK
kb_reference=CRJ2k_REF
```

These last few parameters identify the files used for the checklists and reference files available to the aircraft.

If you want to create and reference multiple model, panel, sound, and texture directories, you would use the naming convention foldername.extension, where the extension is a unique identifier for a configuration set. To refer to the folder in the aircraft.cfg file, you specify the extension. For example, if you had a special IRF panel, you would place it in a folder named panel.ifr and in the aircraft.cfg file the panel parameter would be panel=ifr. If a parameter isn't explicitly set, it automatically refers to the default folder.

The **[forcefeedback]** section contains the settings for force feedback joysticks. The settings adjust the amount of feedback are explained in the Flight Simulator SDK which is usually available from Microsoft's flight simulator web site or TecPilot. These settings allow you to have a different 'feel' customized for each aircraft.

The last sections for this aircraft.cfg file are a number of blank **[fltsim.n]** sections for adding additional versions of the various options for the aircraft.

There are many more parameters available for use in the aircraft.cfg file. A complete list can be found in the Flight Simulator SDK.

The next files in the list, **crj2k_check.txt** and **crj2k_ref.txt**, are text files that contain the checklists and knee board reference information. These files can be customised by using Notepad or any text editor.

The flight dynamics for the aircraft are contained in the **crj5kend.air** file. Every aircraft must have an .air file. The file is in binary format and making other than simple changes to the .air file with the flight dynamics editor included in Flight Simulator requires some knowledge of aerodynamics, as well as a hex editor or other program designed for editing the .air file.

delcrj2e.gif is a screen shot of the aircraft and is not a required file. **FILE_ID.DIZ** is a description file included in the zip file and is used by some download sites. It is not a required file. The last file, readme.txt, we have discussed previously. Although it is not required, it is a good idea to keep the file for reference purposes.

There are a number of other sections that may appear in an aircraft.cfg file. More information on these sections can be found in the SDK. The sections include:

[flight_tuning] Flight control effectiveness parameters, stability parameters, lift parameters, drag parameters

[weight_and_balance] The weight and centre of gravity of the aircraft can be adjusted through these parameters. The sign convention for positions is positive equals longitudinally forward, laterally to the right, and vertically upward.
The moments of inertia (MOI) can also be modified in this section.

[piston_engine] and **[turboprop_engine]** These parameters allow you to scale the power generated for piston and turboprop engines.

[propeller] and **[jet_engine]** The thrust produced can be modified using the parameters in this section.

[ground_reaction] These parameters define the characteristics of the landing gear.

[electrical] These parameters configure the characteristics of the aircraft's electrical system. Each aircraft has a battery and an alternator or generator for each engine.

[pitot_static] These parameters adjust the lag time of the pitot-static system.

[helicopter] The low_realism_stability_scale parameter scales the stability of the Bell 206B helicopter to make the aircraft easier to fly.

The sub folders

Now let's take a look at the contents of the folders. The first is the model folder, which contains two files - **crj5kend.MDL** and **model.cfg.**

The **model.cfg** file specifies which visual models (.mdl files) to render during normal flight and during a crash, as well as the visual model color scheme. A model.cfg file has [models] and [colors] parameters.

crj5kend.MDL contains the information used to create the visual model aircraft in flight simulator.

Next in line is the panel folder. The **panel.cfg** file is located in an aircraft's panel folder and defines the characteristics of the aircraft's cockpit, including window settings, view settings, and gauges. We'll take a more in depth look at this folder later when we install a panel.

The sound folder is next in line. The **sound.cfg** file is located in an aircraft's sound folder and defines the sounds to use for that aircraft. We'll take a more in depth look at this folder later in the Sound section of this chapter.

The last folder is the texture folder. An aircraft's textures are defined by the .bmp files in the aircraft's texture folder. They are "projected" onto the aircraft's parts as specified in the aircraft's visual model (.mdl) file.

Flight Simulator texture files are "mipmapped". A mipmapped texture consists of a sequence of images, each of which is a progressively lower resolution. Mipmapping is a way to improve the quality of rendered textures.

To edit a mipmapped texture and maintain the format, you'll need to use an image tool that is capable of handling these extended format bitmaps.

Figure 2

A texture can be edited using a simple graphics application, however it will be saved as a standard .bmp file instead of a mipmapped .bmp.

This completes the files and folders found in our CRJ installation package. Since everything is in order, we are now ready to actually install the aircraft in Flight Simulator.

The easiest way to move the files and folders from their temporary location to their spot in Flight Simulator is to open Windows Explorer and navigate to your temporary folder. Then click delcrj2e to highlight it, right click and chose "Copy". **[Figure 2]**

Figure 3

Next, navigate to your Flight Simulator folder. If you used the default path for installation settings, this will be C:\Program Files\Microsoft Games\Flight Simulator 2000. One level below the Flight Simulator folder, you will see the \aircraft folder. Right click to highlight Aircraft and then click "Paste". **[Figure 3]** This will create a copy of the delcrj2e folder, including all its files and sub folders. **[Figure 4]** You can now go back to the temporary location of delcrj2e folder and delete it.

An alternative way to install the aircraft is to click and drag the delcrj2e folder to the Aircraft folder and then, while holding down the Shift key, drop it on the Aircraft folder.

Figure 4

We are now ready to start Flight Simulator and go for a flight in our newly installed CRJ 200.

Installing Downloads Into Flight Simulator

Panels

In Flight Simulator, all aircraft-specific files, except for the gauges used by panels, are located in the Aircraft folder. As we saw in the previous chapter, one of the sub folders of the specific aircraft folder is the panel folder. The CRJ we just installed does not have its own specific panel and uses a technique called aliasing in order to use an existing panel. An explanation of aliasing follows the section on sound files.

If you look in the panel folder under the delcrj2e in Flight Simulator, you will find a single file named panel.cfg. If you open that file with a text editor, such as Notepad, you will find the following:

```
[fltsim]
alias=FSFSConv\panel.Jet.Heavy.2
```

When the CRJ is loaded, Flight Simulator looks in the panel.cfg file for the correct panel. The alias parameter points to the generic Flight Simulator panel for a two engine jet aircraft.

While this panel will work fine with the CRJ, in this section we shall install a new one that more closely resembles a real CRJ panel. The one we shall use is still a bit basic but this exercise is designed give you a good working knowledge of the <u>correct</u> way to perform a decent panel installation. You can always update it later for a more complex one.

You should now unzip the second file we downloaded previously, cl65panl.zip. Unzip the file into your temporary folder and then we will take a look at the files and then talk about some specifics on panel files.

After opening the temporary folder where you unzipped the file, you will see six files. **[Figure 1]** The first, CL-65 **PANELREADME.txt**, is not a required file and gives a little information about the author and the restrictions on using the panel. The next file, **CL65Panel.gif**, is a screen shot of the panel and is also not a required file. **CL65PANELFILE_ID.txt** is not a required file as well and is a brief description of the panel. CL-65PanLarge.jpg is a graphic of a close-up of the panel and is not required.

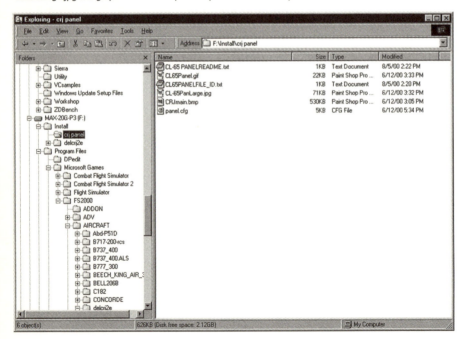

Figure 1

That leaves us with the two required files for this panel - **CRJmain.bmp** and **panel.cfg**. **CRJmain.bmp** is the background graphic used in Flight Simulator to create the panel. The gauges will be overlaid on this background by Flight Simulator to create the finished panel. There are no custom gauges used by this panel. It uses only the default gauges that are installed by Flight Simulator.

Before we actually install the panel, let's take a look at the panel.cfg. The layout of the panel.cfg file is similar to the aircraft file in that it has a number of sections and multiple parameters within each section. Although we are mainly interested in installing a panel, an understanding of what is contained in the panel.cfg file is very beneficial for troubleshooting problems, or if you are interested in designing your own panels. We shall go through each section and briefly discuss its purpose.

If you have more than one version of an aircraft or want to have multiple panels for an aircraft, you will have more than one panel folder. The name of the default panel folder is panel and any additional panel folders must have the name panel.xxx, where xxx is the identifying key we discussed in the aircraft.cfg section.

Each panel configuration has the following individual sections:

```
[Window Titles] section
[Views] section
[WindowXX] section
[Fixed WindowXX] section
[Default View] section
```

[Window Titles] section

The panel system creates the panel windows by using the [Window Titles]
section. The variables in this section are used to set the initial state of each panel window, and each interior view.

For each WindowXX variable in this section, the system creates a panel window. It starts with Window00 and creates panel windows until it reaches the maximum of 63 or there is a break in the numbering.

The panel system creates Fixed Window interior views with Fixed Window30 and continues up through Fixed Window 37. The order in which you assign panel or fixed windows doesn't matter, so long as you start Window00 and Fixed Window30.

When the panel window is created, the system looks for corresponding [WindowXX] and [Fixed WindowXX] sections later in the configuration file. After the panel window is created, the gauges listed in the [WindowXX] section are loaded.

For example, the panel.cfg file for the Cessna 182S contains the following information in this section:

```
[Window Titles]
Window00=Main Panel
Window01=Radio Stack
Window02=GPS
Window03=Annunciator
Window04=Compass
Fixed Window30=Forward right
Fixed Window31=Right
Fixed Window32=Rear Right
```

The string assigned to each WindowXX or Fixed WindowXX is the title displayed for the panel window when it's undocked. It is also displayed as in the list when you click Instrument Panel on the Views menu.

[Views] section

This section specifies the view direction in which panels appear in the aircraft.

Installing Panels into Flight Simulator

[WindowXX] section

The gauges are specified by the gauge00 through gauge100 in the [WindowXX] section. The system loads, starting from gauge00 until it reaches gauge100 or until it finds a break in the progression.

Here's the basic format used for each gauge in a panel:

`gaugenn=gaugefile!gaugename, X, Y, W, usage, parameters`

Where:

gaugenn -- indicates the order in which the gauge is loaded.
gaugefile! - is the specific .GAU file in which the gauge is found.
gaugename - is the name of the gauge you assigned in gauge code

The new style Flight Simulator gauges can have multiple gauges in a single gauge (.gau) file. The older style gauges from Flight Simulator 98 and before have only one gauge per gauge file.

X, Y - the X and Y position of the gauge in millimeters relative to the panel background.

W - is the width of the gauge in millimeters.

Usage - not used.

Parameters - these are passed to the gauge as a text string argument.

[Fixed WindowXX] section

This section specifies which bitmap to use in each of the fixed window views available by manipulating the joystick hat or view keys. There is one Fixed WindowXX entry required for each number assigned in the [Views] section.

[Default View] section

This section sets the overall view for the default 3D window. There are four parameters in this section:

X specifies the X position (horizontal).
Y specifies the Y position (vertical).
SIZE_X specifies the width.
SIZE_Y specifies the height.

Placing Gauges on a Panel Background

The [WindowXX] section is where the gauges are placed on the panel background itself. The gauges are positioned by X and Y coordinates, which correspond to the upper left corner of the gauge.

The following will explain some details on placing your gauges by editing panel.cfg. Note that all panel backgrounds that come with Flight Simulator have both a 640 version and a 1024 version. Panel.cfg loads the correct one depending on your screen resolution.

The panel.cfg file for nearly every aircraft in Flight Simulator 2000 sets size_mm to 640. It also sets the window_size_ratio to 1.0. This means that there is a 1-to-a ratio between pixels and millimetres. This means that for references in the panel.cfg file, pixels and millimetres are interchangeable and will also scale correctly for different screen resolutions.

You do not have to use a bmp that is exactly 640 or 1024 pixels wide, although not doing it creates a little more work for you, because you will need to calculate the ratio in order to have pixels and millimetres interchangeable.

Now that you have an idea of some of the settings used in the panel.cfg file, it's time to install the panel for our CRJ. We will install it as a second configuration, so that the original panel will still be available. This is also demonstrate how to have multiple panels for a single aircraft. The same principles apply to having multiple textures or sounds.

First, go to the delcrj2e folder in the Flight Simulator and click on it. Right click in the right hand window and then click new, then folder. For the folder name, enter panel.new. **[Figure 2]**

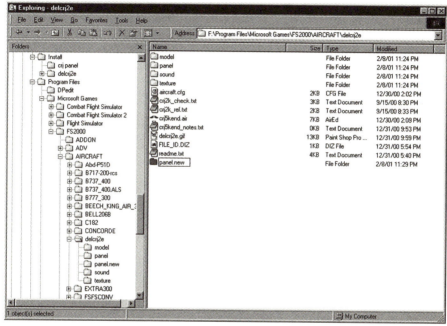

Figure 2

Now copy the contents of the temporary folder where we put the files from cl65panl.zip and paste them into panel.new. You can also highlight the files and drag them to panel.new.

We now have to make a couple of changes in the aircraft.cfg file so that the new panel will show up. First, open the aircraft.cfg file in Notepad. In Notepad, highlight the [fltsim.0] section (shown below), then copy and paste it below the existing section.

```
[fltsim.0]
title=DelawAIR Canadair RJ200-LR
sim=crj5kend
model=
panel=
sound=
texture=
checklists=
kb_checklists=CRJ2k_CHECK
kb_reference=CRJ2k_REF
```

We now need to make three changes in the new [fltsim.0] section. The first is to change [fltsim.0] to [fltsim.1]. The second is to change the name on the title= line. Since this line determines how the aircraft is listed in Flight Simulator, each title needs to be different in order to tell which aircraft is which. The title can be whatever you want. For this tutorial, let's make it DelawAIR Canadair RJ200-LR 2. The third change is to tell Flight Simulator 2000 where to find the panel using the panel= parameter. Change that line to read panel=new .

The [fltsim.1] section should now look like this: **[Figure 3]**

Figure 3

```
[fltsim.1]
title=DelawAIR Canadair RJ200-LR 2
sim=crj5kend
model=
panel=new
sound=
texture=
checklists=
kb_checklists=CRJ2k_CHECK
kb_reference=CRJ2k_REF
```

We have one last change to make. This particular aircraft.cfg file already had numerous blank [fltsim.n] sections, so we need to remove the existing [fltsim.1] section. You will find it after the [forcefeedback] section. Highlight and delete the whole [fltsim.1] section.

Now save the aircraft.cfg file. You are now ready to fire up Flight Simulator and try out the new panel. You should find it in the aircraft list as DelawAIR Canadair RJ200-LR 2 and you will also see the original DelawAIR Canadair RJ200-LR just above it. **[Figure 4]**

Figure 4

If the panel you are installing has custom gauges included, you would copy them to the gauges folder found one level below the main Flight Simulator folder. If you are installing a new panel, any file with an extension of .gau should be placed in this folder. While there should not be an instance where the same name is used for multiple gauges, we have seen it happen. If the two gauges were actually different, this could cause problems with the original panel that uses the gauge in question. In other words, be careful before overwriting gauge files.

We have now completed the conversion of a default panel to a more realistic one for CRJ. Now try flying it and then install a more complex one - perhaps in FS2002!

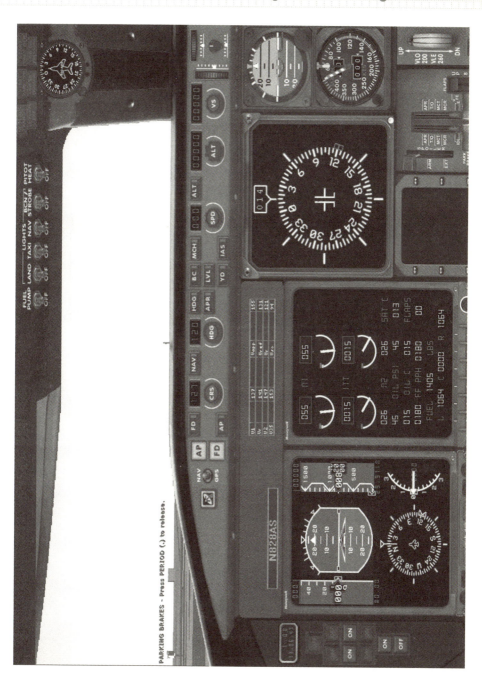

PARKING BRAKES - Press PERIOD (.) to release.

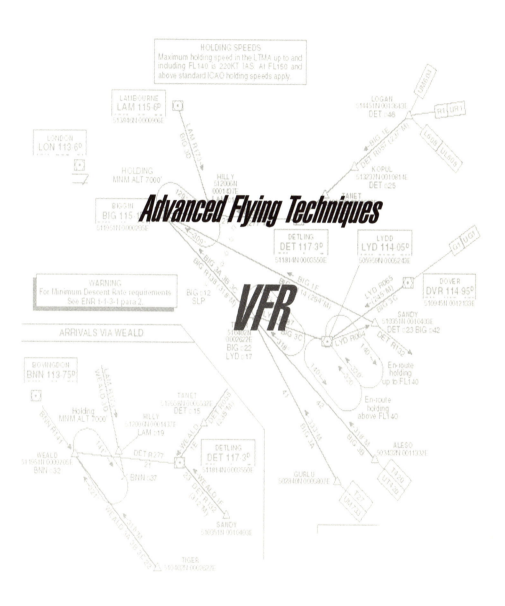

Advanced Flying Techniques
VFR

Advanced Flying Techniques - VFR

What was that the fellow on the radio just said? "Take off at your discretion"? Surely he can do better than that, what happened to "Cleared for departure"? Well alas no, when flying from many small aerodromes, the only person responsible for deciding whether it is safe to take off is you, and you alone. It is a misconception that flight plans are filed for every flight, and that you are continually under the control of "Air Traffic". You are normally controlled at big airports and when airways flying, but for much of the time when flying VFR in the UK you can fly where you want to, without talking to a soul. So the chap talking to you on the radio is saying that he has no objection to you taking off, but it's really up to you. At some aerodromes, he might not even be able to see you or all the runways! So have a good look around for obstacles and other planes and take off at your discretion.

Lets put all this into context. Your task for today is to fly the Cessna 182S from Wellesbourne Mountford runway 05 to Netherthorpe runway 06. If you were to look up Netherthorpe in a flight manual, you would find the magic letters PPR. These stand for Prior Permission Required, which means that you cannot just turn up and land there. So you have telephoned them to get the necessary permission and they are quite happy for you to land there. The weather is quite acceptable for you with the wind is 060 at 5 knots and the visibility at 10 miles, so set this on your flight simulator. While you're at it, add one layer of cumulus clouds with base 2500 feet.

You could take off and head straight for Netherthorpe, calculate the single heading that will take you there and fly it. Being the good pilot you are, you have looked at the map and seen that the straight line route would take you through the controlled airspace at Birmingham and East Midlands Airports. You could still do this, but you would have to get their permission. It's quite possible that they would refuse clearance for you to transit their airspace when you call them on the radio, and you would have to do some rapid re-planning in mid-air. What's more, they may give you permission, vector you all around the sky to avoid other traffic and then leave you with the heart sinking words "resume own navigation". So you have take the sensible choice and planned to fly round the side of their airspace. Your planned height of 2500 feet is good too, as if keeps you under the Birmingham control area and the Lincolnshire Area of Intense Aerial Activity (AIAA), and over the various aerodrome traffic zones along the way.

Your chosen airspeed of 120 knots is well selected as, even when you have adjusted for height and light winds, you will be covering about 2 miles per minute. This makes any in-flight calculations based on speed, distance and time easy; your distance covered in miles on any leg will be roughly twice the minutes you have been flying it. You've done your planning and produced your basic flight log, which we have copied and shown below.

✈ FLIGHT LOG						
Waypoint	HdgM	Alt	Min	ETA	RTA	ATA
Wellesbourne Mountford	054	2500	9			
Rugby	036	2500	8			
Leicester	035	2500	5			
Melton Mowbray	002	2500	8			
1/2	002	2500	8			
Gamston (Retford)	288	2500	4			
Netherthorpe						

A blank copy of this flight log is available at the end of this tutorial on pages 79. If you are using Flight Simulator copy it onto your kneeboard.

If you want get the most learning from the flight you should not use the autopilot, GPS or navigational radios. If you fly accurately, you should not have any difficulty in completing the route. Many aircraft that are used for training have none of these luxuries and pilots are taught to navigate using their mark two eyeballs, which are trained versions of the ones we are born with. In the same way, if you want your simulation to be realistic, you should never press the pause button. Flying is difficult, and one of the problems is that you cannot stop and sort things out. At best you can circle, but even then you need to be careful that you don't drift downwind. If you do decide to circle, make a note of the time at which you start to circle and the time at which you stop circling. You can then add the time you have spent circling to your Estimated Time of Arrival (ETA) for that leg, giving you a Revised Time of Arrival (RTA).

The Good Flight Simmer's Guide

So you are at the threshold of runway 05 about to take off. If all is clear open the throttle and off you go. As you reach 500 feet above the aerodrome carry out a "left hand turn out" by turning onto heading 032. This will avoid flying over Wellesbourne village, which is prohibited below 2000 feet. As you reach 1000 feet turn left again onto 240 and keep climbing. When you reach 2500 feet turn left again to fly directly towards the aerodrome. Reduce engine RPM to 2400 and bring back the throttle to give an indicated airspeed of 120 knots. Everything should be steady by the time you fly over the airfield, so turn onto heading 054 as you pass overhead, keeping at your planned altitude. Make a note of the time and write the minute's part in the Actual Time of Arrival (ATA) column for Wellesbourne Mountford. Hence if you passed overhead Wellesbourne at 14:16 you would write 16 in the ATA column. Now add the 9 minutes leg time and write the minutes part in the ETA column. So using the same example you would write 25 in the ETA column, which is the 16 ATA plus the 9 minutes leg time.

Every time you fly, you should carry out a FREDA check regularly throughout the flight. If you are flying on instruments or at night you should add an 'I' to the end and check for Icing on the airframe and windows.

Now fly your planned heading at your planned altitude, and yes, you do need to fly to the nearest degree. It's not so critical on short legs, but on the longer leg to Gamston, being out by 5 degrees would put you 2 ½ miles to the side of your way-point, and you could easily miss it in poor weather. You also need to maintain your height because the wind will change at different heights, so flying at the wrong height will be the same as flying the wrong heading. It is quite possible that a pilot who flies lower to keep out of cloud will get lost; this is because the change in wind speed will change the actual track flown across the ground. As the minutes on your clock approach your ETA, or RTA if you have one, (i.e. approach 25 using the example time above) start looking for Rugby. You should be able to positively identify Rugby by Draycott Water, and it should be slightly to the right of your track

Once you are established in the cruise, carry out a FREDA check. **FREDA** is an acronym.	
Fuel	Check that you still have enough to reach your destination and your planned alternates, that you are not using the fuel faster than expected, that you are on the correct fuel tank and that the fuel pump is set correctly. The fuel pump is normally off above 1500 feet and on below it.
Radio	Check that you are on the correct frequencies and spoken to everyone you should have. If you were using radio aids, and we are not, you would also check that you had idented all the ones you were using.
Engine	Check that you have set the engine correctly, RPM, manifold pressure, fuel flows, etc. Check all your engine instruments, and use the carburettor heat for about 30 seconds if you have one.
DI	Check that your direction indicator is set to the same as your magnetic compass.
Altimeter	Check that your altimeter is set to the correct setting.

Draycott Water just before Rugby is the easiest way to confirm that you are nearing the first waypoint.

just before you reach Rugby. When you are precisely over the centre of Rugby, write down your ATA and turn onto your planned heading for the next leg. Work out and write down your ETA for Leicester exactly as before by adding the leg time to the Rugby ATA.

Leicester aerodrome is 469 feet AMSL (Above Mean Sea Level) and its aerodrome traffic zone reaches 2000 feet above this. This means that you are flying 31 feet above their zone and don't actually need to talk to them, but it would be extremely wise to call them on the radio to let then know what you are doing. Traffic arriving at Leicester for overhead joins, which is a common arrival procedure, are likely to be in precisely the same bit of airspace as you. So tune 122.125 on your radio, if you could call them, you would say something along the lines of "Leicester Radio, Golf Bravo Sierra Mike Hotel, Cessna One Eight Two routing Wellesbourne to Netherthorpe, 2,500 feet QNH 1013, VMC, estimate your overhead at three three, request traffic information". I dream of the days when flight simulators can hear and respond. VMC is Visual Meteorological Conditions and tells them you can see where you are going. QNH 1013 tells him the setting on your altimeter (1013 is the UK equivalent of 29.9) and that you think you are flying at 2500 AMSL. When you reach Leicester Aerodrome, positively identify it. It is just to the East of Leicester and has a distinctive triangular runway pattern with two grass strips nestled in the middle. Again, when you are exactly overhead the airfield, note the ATA, calculate the next ETA and turn onto the new heading to Melton Mowbray.

The next leg of your journey is only 5 minutes, but how will you recognise Melton Mowbray? Most towns of a reasonable size have some unique features that can be identified from the air. They don't have to be unique in the world, just in the area you are flying. As you are flying so accurately, there will normally be little doubt about where you are, but if the direction indicator has drifted, and they do, you could be a long way off track. Therefore, as we have done before, it is essential to positively identify your position as often as possible. In the case of Melton Mowbray, it is medium size, on a river running East to West, and has several roads and railways (including disused ones) running into it. The river and

The layout of the runways at Leicester Aerodrome are very distinctive.

most of the roads are visible in FS, so you should know when you get there. What's more, the leg is so short that if you have been doing your FREDA checks properly, the town that appears after 5 minutes can only be Melton Mowbray. Also if you flight simulator does not have Melton Mowbray, and you are flying accurately, you could probably get away with just turning after 5 minutes.

The leg to Gamston is quite long and so it is sensible to split it into two halves.

The mid point is neatly just past the grass strips of Newton aerodrome and almost over the only significant lakes on your track. You will be flying just under the Lincolnshire AIAA and about 10 miles to the West of a whole gamut of military airbases. They are rather keen that you don't stray into their busy airspace and this is one reason that they offer a LARS (Lower Airspace Radar Service). So put Waddington 127.35 on your comms radio and say hello to them too. As they like to pick you up on their radar, your call this time would be something like

Melton Mowbray is not so easy to identify, but if you are flying well and the time is right, this must be your way point

"Waddington, Golf Bravo Sierra Mike Hotel, Cessna One Eight Two, overhead Melton Mowbray, heading zero zero two, estimate Gamston at five four, 2,500 feet QNH 1013, VMC, destination Netherthorpe, request traffic information"

By now you will have the hang of this tutorial, so the turn and checks at Gamston will be easy. Don't forget to give Gamston Radio a call on 130.475 before you get there to let them know what you're doing, and then change again to Netherthorpe Radio on 123.275. The grass strips of Netherthorpe airfield will soon pop into view. It is 250 feet AMSL, so set your altimeter to the QFE of 29.65 or 1005. Descend into a right hand downwind position for runway 06. That will be about 2 miles to the South East of the field at a circuit height of

Approaching Gamston / Retford. Don't always expect your waypoints to pop up out of the haze exactly in front of you.

800 feet, heading 240. Quickly carry out your pre-landing checks; most importantly that the Mixture and Propeller levers are fully forward, and the fuel pump is on. Don't forget to put your gear down if you are flying a retractable. There are only 407 metres of runway available for you to land on, so you will have to use your short field technique. You will need to hold a low approach speed of 65 knots, give yourself a long final approach, not get too low too soon and land just past the threshold. As you fly your approach, keep the picture of the runway steady in the windscreen; the shape of the runway should stay the same and just getting bigger.

If you can't find Netherthorpe then you will have to fly to your alternate airport. Gamston or even Waddington would be suitable, but there are plenty of others around to the East of Netherthorpe.

You should use the method given above for all your cross-country flying. Start by splitting your route (using a map, aviation or otherwise) into short sectors between easily identifiable geographical features. Measure the true headings, allow for the wind and then add magnetic variation (use about 4 degrees in the UK) to get the magnetic headings you will fly. Measure the leg distances between the waypoints and work out the time per leg. Then all you have to do is to fly your plan accurately logging what you do as you go along. If you have trouble with the calculations, you can start by setting the wind to zero and flying at 120 knots, then the time in min-

Just before touchdown, if you have difficulty stopping, try a slower approach.

utes for each leg will be half the distance. Hence if you have measured the distance as sixteen miles, plan the time for this leg as 8 minutes.

The shape of the runway stays the same, it just gets bigger as you get closer.

When you are relaxed with the flight from Wellesbourne Mountford to Netherthorpe, try making the visibility worse. A newly qualified Private Pilot in the UK can fly in visibilities down to 5 miles, so why not try that. Once you have mastered this navigation method you should be able to find your way to most airfields in the World. All you have to do is to fly accurately enough. Just remember the three golden rules of navigation: don't get lost, don't get lost and don't get lost.

✈ FLIGHT LOG

Waypoint	HdgM	Alt	Min	ETA	RTA	ATA
Wellesbourne Mountford	054	2500	9			
Rugby	036	2500	8			
Leicester	035	2500	5			
Melton Mowbray	002	2500	8			
1/2	002	2500	8			
Gamston (Retford)	288	2500	4			
Netherthorpe						

✈ FLIGHT LOG - VFR

Waypoint	HdgM	Alt	Min	ETA	RTA	ATA

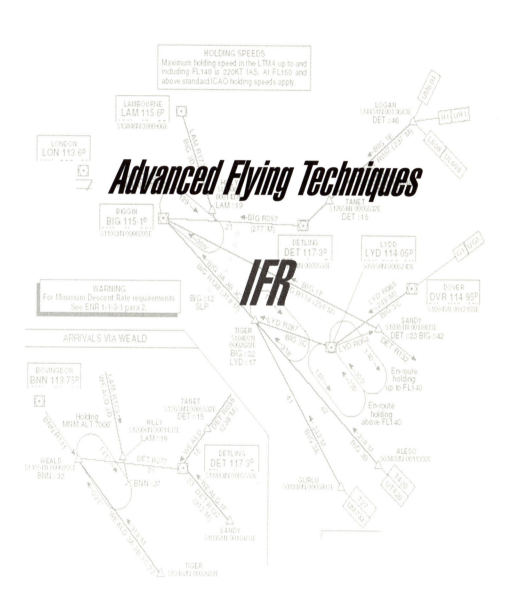

Advanced Flying Techniques

IFR

The Non-Directional Beacon is in many ways the easiest beacon to use for navigation when flying. Keep the aircraft pointing at the NDB and you will eventually reach it. Unfortunately it is also one of the hardest beacons to fly properly because it gives no distance information and because there will always be wind blowing you off course. It gets harder still, because you are also likely to spend as much time flying away from a beacon as you are flying towards one, and flying away from an NDB on a specific radial can be tricky. The route selected for this tutorial is therefore easy to fly badly but difficult to fly properly and accurately.

The flight is a fairly straightforward beacon hop from Exeter to the Scilly Isles, with all the navigation done using NDB's. To add a bit of spice at the end of the flight, the instrument approach to St Mary's will be an NDB approach to Runway 33. For extra realism you will also be flying the route in some pretty rotten weather, just to give you a reason not to simply follow the coast and then dead reckon across the final leg.

Before dashing to your flight simulator, you will need to know some key facts about the NDB. Firstly the Non-Directional Beacon is, as its name suggests, a beacon that radiates its signal equally in all directions. The instrument that detects it is the Automatic Direction Finder. The needle on the ADF dial points towards the NDB. Well it would if the dial was located on the floor of the cockpit, but you couldn't read it there, so it is moved to the instrument panel. When an NDB is directly in front of the

plane, the needle on the ADF points upwards, when the beacon is to your left the needle points left and so on. What it shows is a relative bearing, which if added to your heading will give the magnetic heading (QDM) to the NDB. If you have idented the beacon successfully as you always should, you can be reasonably confident that the beacon and your instrument are working properly. However, just to be sure, you should tune it to a local NDB before you taxi, and make sure that the needle keeps pointing at the beacon as you taxi round the corners to the runway. Easy.

INITIAL SET UP

Location: EGTE - Exeter
Runway: 08
Coms Radio: 119.8
Nav Radio: Off
ADF Radio: 337 EX
XPDR: 7000
Magnetic Compass: Visible
Pre-takeoff checks: Complete

Weather Setup

Wind direction: 010 True
Wind speed: 10 kts
Cloud base: Overcast at 700 ft
Cloud tops: 8000 ft
Precipitation: Heavy Rain
Visibility: 1 mile

As the ADF needle points towards the beacon, if you roll the plane onto it's side, there will be an error in the needle position. To illustrate this, consider a plane flying one mile high and one mile from the beacon. If you roll plane 90 degrees onto its wingtip, the beacon will now be 45 degrees to the side relative to the plane. The ADF does not care whether the plane is the right way up or not, it simply points to the NDB. This is not a normal instrument manoeuvre, but a much smaller version of the same effect can result in 5 or 10 degrees error during instrument turns, which can sometimes be confusing.

The radio waves from the NDB can also give rise to errors because they don't always travel in straight lines. They are bent by things like coastlines, hills and thunderstorms. This isn't really something you need to worry about too much, because your margins for error should always be big enough for this not to jeopardise your flight safety.

There are a number of settings on the NDB radio and you will need to set your receiver to the right one. The main settings vary from box to box, but the main ones are ADF, ANT, BFO and TEST. ADF is the one to use pretty much all the time in the UK. ANT is for identifying beacons, BFO is for old style NDB's and TEST will displace the needle away from the NDB so that you can check that it re-acquires the beacon properly.

Enough of the theory, the time has come to put it into practise. Position yourself at the beginning of Exeter's Runway 08 in a Cessna 182 RG or something similar. Set the weather given in the sidebar and check that you have your radios set for the first stage of the flight. The tower has just called to say that you are "Clear Takeoff right turn out, surface wind 010 10 knots". So acknowledge the call with "G-BSMH is clear take off", open the throttle and start your journey. As you accelerate down the runway, check that all the engine instruments are giving sensible readings and rotate when you reach your chosen take off speed. Stay in line with the runway extended centreline and at 200 feet raise your undercarriage and then throttle back to 24 inches on the manifold gauge and carry out another check of your engine instruments. You should be tracking towards the Echo X-ray NDB on track 082 degrees, but you will have little time to make any meaningful corrections, so sticking to your calculated heading of 075 will be good enough. Settle down to a 500 feet per minute climb and trim for hands off flight, but do keep your hands on the controls.

✈ FLIGHT LOG

7 Waypoints

Position	ID/Freq		ETA	RTA	ATA	TTLT	GDTG
ALT/MSA	AWY	TRM	HDM	IAS	TIM	W/V	DST
Exeter	EGTE					0	135
4.0 / 3.4		082	078	120 Kts	2	010/10	5
Exeter	EX337					2	131
4.0 / 3.4		243	246	120 Kts	17	010/10	37
Plymouth	PY396.5					19	94
4.0 / 3.0		279	283	120Kts	12	010/10	34
St Mawgan	SM356.5					35	60
4.0 / 2.0		233	236	120Kts	12	010/10	28
Penzance	PH333					47	32
4.0 / 2.0		254	258	120Kts	15	010/10	32
St Mary's	STM321					62	
				120Kts		010/10	0
Scilly Isles	EGHE					62	
				120Kts		010/10	

At 500 feet make a rate one turn to the right to intercept the 243 radial from the EX. This is your first problem. Picking up an outbound radial from a beacon. As luck would not normally have it, this one is quite straightforward, because your rate one turn onto your calculated heading of 246 (the 243 radial plus 3 degrees drift) puts you pretty much on track. You should only have to make fine adjustments. As soon as you are on your planned track, note the ATA on the second line of your plan, add this to the leg time of 17 minutes given in TIM below it to give your ETA for the Pappa Yankee NDB, and write this in the space provided. (There is a blank Flight Log for you to use on page 87)

Getting onto an NDB radial

1. Tracking TO an NDB on radial 060

You find that you have drifted off to the left of track. You are on heading but the needle is pointing to the right. Your QDM is 070.

You turn Right 20 degrees onto 080 to get back onto track. Note that the ADF is now pointing to the left (other side) of the vertical. Your QDM is still 070.

When the QDM is 060 you are back on the radial. Superimpose the ADF needle on top of the DI to see your QDM is 060. Note that you are steering 20 degrees right of your radial and the needle is now 20 degrees left; this is the quickest way to see you are back onto track.

Turn onto the heading of the radial 060 to check you are on the track. The needle is at the top, so you are. As you get better at tracking you will miss out this step, but it is always a handy and quick check.

Now allow for some drift. Here we have allowed 5 degrees drift from the right. Notice that the QDM is still 060.

Continued on page (85)

Tracking away from the Echo X-Ray NDB

The ADF needle will be pointing down which means that the NDB is behind you. It should actually be slightly to the right of this by the 3 degrees that you have calculated in your plan as drift. If it is not you will have to do something about it. To start to get the hang of how to track NDB's, we will do this the long winded, but clearer way. Firstly turn onto the heading of the radial you want to track, that is 243. If the downward pointing ADF needle is to the right, then you are to the left of track and vice-versa. If you are bang on track, it would be sensible to turn right for a little to give you the practice of regaining the NDB radial. To get back onto track you will have to make the needle move even further away from directly downwards, i.e. make the ADF error look worse. So if the needle is to the left, turn left, and the needle moves even more left. It is impossible to say how far to turn because this depends on how far off track you are, but twice the error is a good start point. Also keep your new heading a multiple of 5 or 10 degrees from your chosen track. This will make life simple for you and avoid too much mental gymnastics. So, as an example, if you are flying heading 243 and the ADF shows a relative bearing of 187 degrees, you are to the right of track by 7 degrees, turn left onto 223 and this will bring you back onto your planned radial.

You are now steadily moving back onto track, but how will you know when you are actually on your chosen radial? Well as you move back onto track the needle will slowly move back towards the downwards position. If the needle doesn't move downward, then turn another 10 degrees left, but be patient. It will take more time for the needle to move when you are further away from the beacon. When the offset of the needle from the downwards position is equal to the amount you turned away from your planned heading, you will be back on the radial. So if you chose the 20 degree offset (turning onto 223 degrees) given above, when the ADF shows 20 degrees left of bottom, or 200 degrees relative bearing, you will be back on track. This is easily shown by adding your heading to the relative bearing on the ADF dial. So heading 223 plus relative bearing 200 degrees equals 423 degrees, which is a full circle 360 plus 063. The QDM is 063 and therefore the radial is 180 plus 063, which is our planned track of 243. Now that's far too much maths to do on the ground, let alone in the air, so stick with the method of adding or subtracting say 20 degrees from your track and waiting for the needle to come back into 20 degrees from the downward position.

The RMI, or Radio Magnetic Indicator is an alternative system built in to some aircraft. On these, the ADF heading card is driven to shows the actual heading of the plane, rather than having 360 always at the top. They also have a needle that will point to a VOR in the same way. Once you understand how RMI's are used, these make life easier, but until you understand how to use them properly, stick to the smaller planes with the basic instrument. The standard ADF has a manually controlled heading card and if you set your actual heading on this, you can read the beacon QDM directly from the ADF needle. This is great if you are flying one heading for a long time and looking for intercepts rather than tracking heading, however it will tend to confuse if you are tracking the beacon itself. You are best to leave the heading card with 360 at the top for most of your navigational needs.

Having successfully intercepted the track you should now fly your planned heading and see how good it is. If you drift off to one side, get back on track, using the technique you have just learnt, and then choose a new heading to steer which you think will keep you on track. So if you drifted to the left, regain your track and choose a new heading to follow which is slightly right. Again make life easy for yourself and try an easy heading to start with.

When you are half way to Plymouth, about 9 minutes before your ETA, set Plymouth Approach 133.55 on your Coms radio and the PY NDB 396.5 on your ADF and ident it. Now is a good time to do a FREDAI check. Call up Plymouth and give them your details "Plymouth Approach this is G-BSMH 18 miles North East of Papa Yankee, 4,000 feet, IMC, Estimate Papa Yankee at 45 (or whatever your ETA is), request traffic information". Sadly they won't reply.

Tracking towards the Papa Yankee NDB

Once you have tuned in the new NDB, you will find that the needle swings round to near the vertically up position. It is unlikely that you will be exactly on track, which is the 243 radial to PY, so you will again have to manoeuvre to get back onto your planned course. The good thing is that tracking towards an NDB is generally easier than tracking away from one. So if you cracked the last half of this sector, you will find this half much easier. Again we will use a method that avoids too much cerebral aerobatics.

Your planned heading, allowing for wind was 246, but you have probably had to adjust this during the last stage. Whatever you have done, if your QDM to the beacon is 243, simply stay on your heading and monitor the situation. If your luck is not so good, or you can't work out the QDM, turn onto heading 243 and look where the ADF needle is. If you are bang on track, the needle will point straight up and all you have to do is to turn back onto the heading you were on one minute ago. However, it is more likely to be pointing to one side or the other, but hopefully only slightly.

As before, if the needle is pointing to the right, you must turn right, and vice versa. But this time when you turn right, the needle moves towards the vertical. To get back on track, you must keep turning until the needle is pointing somewhere on the opposite side of vertical. So if the needle originally was pointing to the right of vertical, you will turn right until the needle is pointing to the left of vertical. How much? Well again it depends on how far you are off track and how far from the beacon. A good starting point is to turn until the ADF needle is as far on the opposite side of vertical as it was on the original side. Again, make this a good round number. So as an example, if the needle is 12 degrees to the right, steering 24 degrees right onto 267 would make the needle show 12 degrees left. In this case 25 degrees is an easier figure to work with, so it would be better to turn right onto heading 268.

This will bring you back onto track, but how do you know when you are there this time? Once your QDM is the required track you are on track, but again this requires mental agility during times of stress. The easy way is to remember how much you turned away from your desired track, in our example this was 25 degrees, and wait until the ADF needle shows this displacement from the vertical.

Once you are back on track, you should select a new heading that accounts for the wind. As you have just changed to the new beacon and are attempting to use the same track, the final heading you had when tracking away from Echo X-Ray will probably be a good initial heading.

A good understanding of these methods for tracking to and from NDB's is absolutely fundamental to their use, so you need to be absolutely certain you understand how it all works.

Before we move on, a word of warning about tracking towards an NDB. If you turn until the needle is pointing vertically upwards and simply keep it there, you will eventually reach the beacon. Although this will get you to your destination, it comes with a big health warning and a few severe black marks in any flying test. The problem is that you will drift further and further downwind, and in strong winds this could push you into a dangerous area you have not planned for. Crashing during a flight test is usually one of those points that examiners are quite picky about.

Now you are firmly established on your radial into the Papa Yankee, allowing for the wind and flying on a steady heading. You should be able to track along this radial until you pass directly over the beacon, when the needle will suddenly swing through 180 degrees to point more or less downwards. All will probably go smoothly until you start to get close to the beacon, when things may begin to get out of control. What is happening is that as you approach the beacon, your margin for error is dramatically reduced. At 15 miles ½ mile off track is 2 degrees on the ADF, which is barely noticeable, whereas at 1 mile the same error is 26 degrees. To be within 2 degrees at 1 mile you must be within 200 feet of the track. The consequence of this is that you have to make fine adjust-

2. Tracking FROM a beacon on radial 060

Again you find that you have drifted off to the left of your track. You are on the correct heading but the needle is pointing to the right. The QDM is 040.

You turn Right 30 degrees onto 270 to get back onto track. Note that the ADF is now pointing even further to the right of the vertical. It actually looks as if things have got worse, but they have not. Your QDM is still 040.

When the QDM is 060 you are back on the radial. As before superimpose the ADF needle on top of the DI to see your QDM is 060. Note that you are steering 30 degrees right and the needle is 30 degrees right; again this is the quickest way to see you are back onto track.

Turn onto heading 240 to check that you are on the 060 radial. The needle is at the bottom because you are tracking from the beacon. Again you will miss out this step when you have cracked the technique.

Allow for some drift. This time we have allowed 10 degrees drift from the right and again the QDM is remains as 060.

ments as you approach the beacon and constantly keep on top of the situation. When you get very close to the beacon, you are probably best to stay on your heading and accept it if you are a little to the side, so long as this is measured in feet not miles. Your closeness to the beacon can be judged by a combination of your ETA and the twitchiness of the needle.

So that's it. You've cracked the ADF. All you have to do on your flight is to follow your plan through St Mawgan and Penzance, using the piloting skills described above. You will be an old hand at NDB tracking by the time you get to Penzance and are heading for St Mary's.

St Mary's Approach

Unfortunately, the weather in the Scilly Isles is no better than it was in Exeter, which means that you will have to fly an NDB approach, so there's a bit of fun still to come. The approach we are going to fly is the NDB Rwy 33. If you study the approach plate, you will see that the key elements of this approach are:

- Arrive at the beacon at 1500 feet QNH and your chosen approach speed. 90 knots would be suitable.
- Turn outbound on the 170 degrees NDB radial and start your stopwatch.
- Descend gently to 1420 feet QNH.
- At 2 ½ minutes, make a left hand rate one turn to intercept the 335 degree inbound radial.
- Once your QDM to the beacon is within 5 degrees of the 335 radial, set QFE and descend to 404 feet.

Don't be worried about this, because it is simply a case of NDB tracking outbound and then inbound again, exactly as you have done all the way from Exeter. The only differences are that you are closer to the beacon, which makes everything more sensitive, and that you have relatively little time to get things sorted out.

The hardest part of this is to quickly latch on to your inbound 335 radial. The best trick for this is to use the red 45 degrees marks on your Direction Indicator. What you are looking for is a QDM of 335 at the same time as having 45 degrees left to turn. So while you are turning left onto 335, watch for this heading to pass the red 45 degree mark. Exactly at this moment, look to see where the ADF needle is. If it shows 45 degrees to the left, things are going well, so just continue. If the ADF needle nearer to the upright position, roll out of the turn on 010, and wait until the ADF does show 45 degrees left before completing the turn onto final approach. If on the other hand the ADF shows more than 45 degrees left, you will either have to increase your turn rate slightly, or keep turning past 335 to recapture the radial.

If it all goes horribly wrong, remember that you must follow the published "Missed Approach" procedure and climb ahead on 335 to 1500 feet and then turn left back to the beacon and start again.

The VOR and ILS seem to be the favourite navaids these days because of their precision. However, NDB's are still used throughout the world and you will find it highly satisfying to fly an excellent NDB approach and find the runway exactly where you expected it to be. If you practise using this flight a few times you will have no difficulty at all executing a perfect approach, which will doubtless be followed by a landing as smooth as silk.

✈ FLIGHT LOG - IFR

Position	ID/Freq		ETA	RTA	ATA	TTLT	GDTG
ALT/MSA	AWY	TRM	HDM	IAS	TIM	W/V	DST

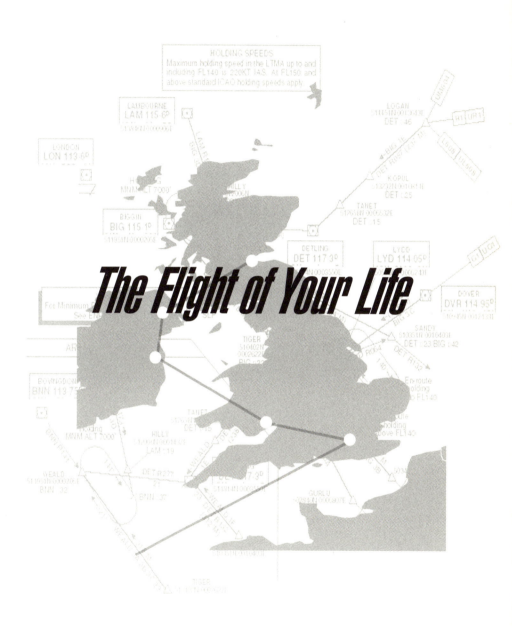

The Flight of Your Life

The Flight of your Life

We decided you should undertake a challenge of proportionate size and complex enough to keep you busy for a good length of time. This flight will take close to a full working day to complete. But don't think that you will be able to take off, point the nose at the destination and then sit back for a few hours watching the scenery roll away. You will be kept pretty busy throughout the flight and because the it is based on the real flight environment, those helpful Air Traffic Control people will occasionally tell you to divert from your planned route for things like traffic avoidance.

There are five different sections to this monster flight that will take you from the Channel Islands to another four major airports in Britain and Ireland. The challenge has not been made specific to any flight simulator, which means that you should be able to complete the flight on any good simulator that has the major airports and navigational aids for the British Isles. The route has been planned in sections, with a flight log for each, so you can either fly it as one large adventure or fly it as a series of shorter flights. If you have an aviation map, you may find it helpful to keep it close by. If not, try to monitor your position on an ordinary atlas. Although neither is necessary to complete this flight, you will doubtless find that some kind of map will help you orientate yourself and keep abreast of roughly where you are.

The route for the flight will take you from Jersey to Edinburgh via Gatwick, Cardiff, Dublin and Belfast, with a landing at each. Most of the navigation will be using radio aids, although you will not always be flying directly to and from them. The choice of plane is also up to you, however it must be capable of cruising at 120 knots and climbing at 500 feet per minute at 90 knots. You will also need a plane that can fly a sensible approach at 75 knots. This probably means that it is one of the light singles found on your simulator, such as a Cessna 182.

Don't worry about whether it's day or night either. A number of the waypoints are underneath the airways reporting points and this means that they are not necessarily related to any specific ground feature. Daytime may help you to keep tabs on your location, but is not essential. The same applies to the weather, select whatever you prefer. If you like zooming in an out of clouds, then set it up with your favourite conditions. The occasional reporting point has a ground feature, so you are probably best to stick to scattered cloud with a reasonable cloud base. You will need to set your cloud-base to give you enough sky around the airports for visual approach and landings. You also should set the wind to 130 degrees 5 knots and leave the atmospheric barometric pressure at 1013 or 29.9 so that Altitude and Flight Levels are equal.

The only other addition to realism is that the pause facility of your flight simulator should never be used; whoever heard of a plane with a pause button. The way to avoid needing to use the pause button is to always stay "ahead of the aeroplane". This means that you should:

⇒ Know everything you are going to do before you reach the place you have to do it. What heading will I change to? Will I turn left or right? What Altitude will I be at? Which side will the cross wind be from when I am landing?

⇒ Do everything as soon as you sensibly can, so that things are ready when you need them. Can I change frequency now? Can I do my FREDA check before I get to the waypoint? Can I ident the beacon now?

⇒ Do as much on the ground as you can, before you take off. Have I made a note of all the frequencies I will need? Have I tuned in all the first frequencies and radials? Do I have enough fuel?

The opposite of this is that the plane gets ahead of you, and you will have flown past a waypoint without knowing what direction you should have turned, need a new frequency which is not tuned in and suddenly have to change fuel tanks because the one you are on has run dry....all at the same time. Always stay ahead of the plane!

Each sector of the flight has a "flight log" prepared for you, which may at first seem odd and sparsely filled to you. Some people like to have absolutely everything that they could possibly need on a flight log, but this can make them unwieldy. It can be awkward juggling the flight log and map on your knee in a cockpit, and if they are both kept to about A5, you will find life much easier. The type of flight log used here is at the minimalist end of things, however it does contain the information that you will require on your flight. Have a look at the flight logs and familiarise yourself with their format. They are all in a condensed format that is similar to that used by many commercial operations. Each waypoint has two lines and you'll pick up how to use it as we fly the individual sectors. The cryptic headings are given as the top two lines. If you look at the sector from Dublin to Belfast and the two lines for Garristown you will see that the navigation aid there is an NDB with frequency 407 and ident GAR. How do we know it is an NDB? With this frequency it must be, try tuning 407 onto your NAV radio and you should find that it cannot be a VOR. There is a space for you to insert your Estimated Time of Arrival (ETA), any Revised Time of Arrival (RTA) and your Actual Time of Arrival (ATA). Next come the total time (TTLT) of the flight so far and the remaining Ground Distance To Go (GDTG). The second line starts with the altitude you plan to fly at and the Minimum Safe Altitude (ALT/MSA). Then comes any airway (AWY), in this case B1 (although you will be flying under it and not in it), the magnetic track (TRM), your expected magnetic heading (HDM) which allows for the wind, the indicated airspeed (IAS) you will fly, the time for the leg (TIM), the wind that was used in planning (W/V) and finally the distance on the leg (DST).

Although we have given you some hints along the way and a little guidance, this flight is intended to be a challenge. We have therefore not given you too much guidance and have left plenty for you to work out for yourself. We hope that all of you will be able to fly the route and that only your accuracy in flying it will differ.

One final point before we move to the actual flying, don't forget to refuel when you need to. Many simulated flights are very short by comparison to the fuel tank capacity in the planes, which means that you never need to fill up. That's not the case here. You may scrape home without filling up, but you would never risk running out of fuel in real life, otherwise your life could become not so real anymore, and you do not have a personal reset button.

Jersey to Gatwick

The time has come to go flying, so set yourself up on the "into wind" runway and review your flight log. The first thing that you should spot is that the airway is given as ORTAC 1B. This is not an airway, but a Standard Instrument Departure or SID. Have a look at the appropriate "Plate" and you will see that the departure takes us directly to the Jersey VOR so tune 112.2 on your NAV radio and ident it as JSY. Turn the OBS on the dial until needle is in the middle and showing "to", keep in your mind that we are going to turn left onto the 006 "from" radial at the VOR (starting the turn just before it) and continue climbing to 5000 feet. You have to be at this altitude by ANGLA and you will have no difficulty achieving this, but you should check it as your DME passes 29 miles from JSY. Stay at 5000 feet, which today is

✈ FLIGHT LOG — 7 Waypoints

Position	ID/Freq		ETA	RTA	ATA	TTLT	GDTG
ALT/MSA	AWY	TRM	HDM	IAS	TIM	W/V	DST
Jersey	EGJJ					0	150
5.0 / 1.9	ORTAC 1B	013	015	120 Kts	22	130 / 05	48
ORTAC 1B	ORTAC112.2					22	102
2.5 / 1.9	R1	043	046	120 Kts	20	130 / 05	40
KATHY						42	61
2.5 / 1.0	R1	044	046	120 Kts	6	130 / 05	13
LUCCO						48	48
2.5 / 1.8	R1	044	046	120 Kts	14	130 /05	28
Midhurst	MID114					62	20
1.5 / 2.0		113	114	120 Kts	4	130 / 05	7
Billingshurst	VRP					65	
1.5 / 1.5		054	056	120 Kts	6	130 / 05	13
Gatwick	EGKK					65	
0.2 / 1.5				120 Kts		130 / 05	

Flight Level 50 until you reach ORTAC, which is DME 47 from JSY. No doubt your trusty first officer is radioing the necessary position reports at ANGLA and ORTAC.

It's time to get airborne. So if you have completed all your pre-flight checks, open the throttles and away you go. As you take off, write the ATA against Jersey on your flight log and fly the SID you have just reviewed.

Climb at 500 fpm and 90 knots until you reach your cruising altitude and then set the engine controls to give you your planned IAS of 120 knots. You now need to insert your ETA for ORTAC and there are two options. The correct way would be to calculate your expected time at ORTAC allowing for your climb, that's not actually too difficult, but you may be will find it easier to simply add the time on route to your Jersey ATA and subsequently enter an RTA once you are level and steady in the cruise. You know that your speed is 120 knots and that ORTAC is 47 DME, so half the distance left to run is about the same as your time in minutes to reach ORTAC.

The flight is going well and you reach ORTAC, filling in your ATA and calculating your ETA for KATHY. You are dropping out of the airway at this point so set yourself up for a 120 knot cruise decent at 500 fpm down to your planned altitude of 2500 feet. The planned track is underneath R1, and to the North of ORTAC this is based on the 044 radial towards Midhurst. So tune 114 on your NAV and ident it. What, no beacon? Well the range of a VOR is limited by the curvature of the Earth, and a rough approximation of the maximum distance at which you will pick up a usable signal is 1.25 times the square root of your altitude. So at 2500 feet you would not expect a signal until about 62 miles. ORTAC is about 82 miles from Midhurst and therefore it is not surprising that you can't receive it yet, if you can, count yourself lucky. Until you do receive it you will have to accurately fly your planned heading.

Fly your way through KATHY and LUCCO keeping a check on your position and times. At this stage you would also no doubt longing for land to appear, just in case you should decide it has done enough for the day. Continue under airway R1 to Midhurst, but tune in Gatwick ATIS on 136.52 for their weather and then ask your co-pilot to contact them before you reach the beacon to request clearance to enter their zone. They clear you to enter the zone at Billingshurst not above 1500 feet and tell you to report entering the zone. Turn right and reset your OBS to 113 and descend fairly promptly to 1500 feet. You need to be on the 113 radial at this height before you reach DME 7 miles, which is the position of the Visual Reporting Point.

Gatwick are very kind to you and tell you to report right base for 08R without any positioning turns or holds, so carry out all your pre-landing checks including the all important check for "three greens", which are the "undercarriage down and locked" lights. As you land note the time and write it in the space for the Gatwick ATA.

Gatwick to Cardiff

Well most of the flight to Gatwick was fairly straightforward and involved beacon hopping from one VOR to the next. Even the Jersey SID was not too demanding. The next stage across to Cardiff is quite a bit more complex as you will be threading your way around Control Areas and Military Air Traffic Zones. You could get clearance to fly straight through them, but many pilots will choose to go around the side, and that is our plan for this flight.

✈ FLIGHT LOG   7 Waypoints

Position	ID/Freq		ETA	RTA	ATA	TTLT	GDTG
ALT/MSA	AWY	TRM	HDM	IAS	TIM	W/V	DST
Gatwick	EGKK					0	138
2.0 / 1.5	ORTAC 1B	235	233	120 Kts	6	130 / 05	13
Billingshurst MID 113/7	VRB					6	125
2.5 / 1.5	R1	272	271	120 kts	9	130 / 05	18
Petersfield						15	107
2.5 / 2.3		327	325	120 kts	13	130 / 05	29
Newbury						28	78
4.0 / 1.5		266	265	120 Kts	28	130 / 05	60
Burnham on Sea						56	18
1.5 / 1.5		332	332	120 Kts	4	130 / 05	10
Flat Holm Lighthouse						60	
1.5 / 1.5		282	280	120 Kts	4	130 / 05	8
Cardiff	EGFF					60	
0.2 / 1.5				120 Kts		130 / 05	

The first leg of your flight is back to the VRP at Billingshurst. Now if you were smart, you would have had a good look at Billingshurst on the way into Gatwick so that you could recognise it on the way out. Whether you did or not, take off from Gatwick 08R, climb ahead to 500 feet above the runway, then turn right onto 170 and climb to 1000 feet above the runway before setting heading for Billingshurst. Be careful because the planned heading is from the centre of the airport and you are now a mile or two to the south of the runway, so you will have to decide the best heading to use to fly directly to the VRP. The common navigation rule of thumb is that for every 60 miles you fly 1 degree off track is equivalent to 1 mile off track. Now you are about one mile to the side of track with about 13 miles to run. Using the 1 in 60 rule gives about 4 or 5 degrees of correction required, so fly to the right of your planned heading by this amount. Check your VOR is still set for MID radial 113 and as you approach Billingshurst, monitor that the VOR needle passes exactly through the centre when the DME is at 7 miles. If it is not going to be right, change your heading accordingly. When you are at Billingshurst, a call is made to Gatwick to tell them that you are leaving their zone.

Over Billingshurst you will need to turn onto your next heading, and this time you are heading for an intercept on a Goodwood VOR radial. So tune in Goodwood on 114.75 and check the ident is GWC, as you always should. You are aiming to intercept the 327 radial at 11.5 miles DME, and as you have just intercepted the MID radial, you know exactly how to do it. Now follow this radial until you reach DME 41 miles. However as this is a long way from the beacon and we want to find the position accurately, tune in the Compton VOR on 114.35, ideally on a second NAV box, and ident it as CPT; your turning point will then be on the "from" radial 232 at 10 miles DME.

The next leg to Burnham-on-Sea, which is on the East bank of the Bristol Channel half way between Bridgewater and Weston Super-Mare, is quite tricky. It is a long leg and there are no suitable nearby beacons to check your track. It would be possible to increase the accuracy of the flight using RNAV, but not many simulators have the Area Navigation equipment that can create virtual VOR's anywhere. You could also get a position fix mid way using a technique that moves NDB's to a virtual position, however the method is somewhat tricky. Nevertheless it would be sensible to get a midpoint fix, and you can get it from the Southampton VOR. Tune this in on 113.35 and ident it as SAM. When you are at the mid point you should be at DME 39 miles on the 310 radial. We have chosen the mid point because this makes any corrections relatively easy.

The leg is 60 mile long and is calculated as taking 28 minutes. The trick is to fly your planned heading to within one degree (yes, one degree) and wait until the needle for the SAM is in the centre. If everything is according to plan, this will happen about 14 minutes after leaving the Newbury turning point. It is more likely that it will be a little early or a little late, but the key thing is the DME. If this is greater than 39 miles you are to the right of track and vice versa. You need to work out how far off track you are. Start by calculating the difference between the actual DME and the planned 39 miles, so if the DME reads 47, the difference is 8. Next, because the SAM 310 radial crosses your track at about 45 degrees, you take three quarters of this to get 6. In this example, over the last 30 miles you would have drifted 6 miles to the right, so using the 1 in 60 rule, you know that you have drifted 12 degrees right of track. (6 in 30 is equal to 12 in 60). This means that something has gone wrong and you should have actually steered 12 degrees left of the heading you flew. You now need to ask yourself how accurately you flew your planned heading. If you were terrible at holding your heading, simply steer the 12 degrees left of 332 onto 320 to take you to your next waypoint, but work much harder at it. The 12 degrees applies equally to the second half of the sector because you are pretty close to half way. If your flying was perfect, something unplanned for has happened, typically the wind would be different to your plan. In this case you need to adjust the 12 degrees that would have kept you on track, plus a further 12 degrees correct the drift to the side and take you to your destination, hence onto 308. This is critical, so make sure you understand the technique. Also the 12 degrees used above is just an example, you will need to use the same method but with your own figures from your instruments during your flight.

If you reach the halfway point before or after your expected 14 minutes, you can also calculate an RTA for Burnham-on-Sea. However you need to be careful because if you are off track as well, your timing will be affected even if your ETA has not changed. This is because the SAM radial crosses your track at an angle.

Once you have reached Burnham-on-Sea, the leg to Flat Holme Lighthouse, which is on Brecon radial 171 at DME 22, should be easy. Tune it in on 117.45 and, as usual, ident the beacon as BCN. It. Of course, you tune the ATIS on 119.475 and contact approach on 125.85 five minutes before the Flat Holme Lighthouse VRP. You are cleared to enter their zone and transferred to the tower on 125.00 who clear you to join downwind for runway 12. There are no delays, so fly your standard approach and land.

✈ FLIGHT LOG — 8 Waypoints

Position	ID/Freq	TRM	ETA	RTA	ATA	TTLT	GDTG
ALT/MSA	AWY	TRM	HDM	IAS	TIM	W/V	DST
Cardiff	EGFF					0	191
4.5 / 3.0		323	323	120Kts	16	130 / 05	36
AMMAN	AMMAN					16	155
4.5 / 3.0	G1	289	288	120Kts	18	130 / 05	40
Strumble	STU113.1					35	115
4.5 / 1.0	G1	294	293	120Kts	14	130 / 05	31
SLANY	SLANY					48	84
4.5 / 1.0		331	331	120Kts	12	130 /05	27
BEPAN	BEPAN					60	57
3.0 / 2.0		026	028	120Kts	13	130 / 05	28
Wicklow Head						73	29
1.5 / 1.7		350	352	120Kts	7	130 / 05	15
BRAY	VRP					80	
0.2 / 0.2		343	344	120Kts	7	130 / 05	15
Dublin	EIDW					80	
						130 / 05	

Cardiff to Dublin

You are "cleared for take off, left hand turn out" and your initial track is to the North West. Air traffic controllers are usually very helpful like this and it means that you can take off and then turn left through 90 degrees at 500 feet above the aerodrome and then onto your track at about 1000 feet while climbing to 4500 feet. Remember to compensate for your distance to the side using the "one in sixty" rule. Your first waypoint is underneath AMMAN on airway G1 and it is 28 DME on the Bravo Charlie Delta 289 radial. As before you will need to make the necessary adjustments along the way to intercept the radial at the correct position. Also, just for the fun of it, see how accurately you can predict your ETA for AMMAN. The flight through Strumble and on to SLANY should be pretty straightforward. To get from SLANY to BEPAN you will have to fly from one VOR DME fix to another. This should be no more difficult that anything you have done so far, all you have to do is to fly exactly what you have on your plan and then make adjustments if you need to. It's worth noting that one of the key principles of navigation is that you always fly to your plan. When things go wrong you modify your plan and fly to the new plan, but you are still fly to your plan, albeit a modified one. You should not fudge your plan, as that is a recipe for getting lost. If you always fly to a plan you will at least know where you should be, and the chances are that you will be not too far away from that place. You may also suddenly realise that you have flown the wrong heading, such as flying 234 when you should have flown 324, but so long as you are keeping your log up to date you can easily work out roughly where you are.

Once you reach BEPAN, you should be over land near Kilmuckridge and your next two legs take you up the coast to Wicklow Head, which is the most prominent headland and should pop into view at about the right time, if your simulator doesn't show it, make the turn at your planned time and carry on to Dublin's Visual Reporting Point at Bray which can also be located by the Dublin VOR DME on 114.9 at 20 miles on the 165 radial.

You've been cleared to enter the Dublin Zone and told to report at Bray. When you call them at Bray, they tell you to turn heading 020, and after 5 minutes they radio you again and tell you to turn left onto 360. Five minutes later you hear your call sign on the radio again, and it is the controller telling you to report left base for runway 16.

While you are tracking to the airfield you have a bit of spare time to carry out your BUMFPICHH checks, which I know you do every time before you land. In case you have forgotten, these are Brakes, Undercarriage, Mixture, Fuel, Propeller pitch, Instruments, Carb heat, Harnesses and Hatches. Check that everything is as it should be. Keeping on the subject of checks, you have been doing your FREDA checks regularly throughout your flight, haven't you. These stand for Fuel, Radio settings, Engine, Direction Indicator matching the magnetic compass and Altimeter set to the correct pressure.

✈ FLIGHT LOG

7 Waypoints

Position	ID/Freq		ETA	RTA	ATA	TTLT	GDTG
ALT/MSA	AWY	TRM	HDM	IAS	TIM	W/V	DST
Dublin	EIDW					0	118
4.5 / 3.0		320	321	120Kts	4	130/05	9
Garristown	GAR407					4	109
4.5 / 3.0	B1	291	290	120Kts	15	130 / 05	32
RANAR	RANAR					19	77
4.5 / 1.0		363	004	120Kts	9	130 / 05	19
Cavan						27	58
4.5 / 1.0		048	050	120Kts	10	130 / 05	21
Monaghan						37	37
3.0 / 2.0		066	068	120Kts	10	130 / 05	21
Portadown	VRP					47	
1.5 / 1.7		036	038	120Kts	8	130 / 05	16
Belfast Aldergrove	EGAA					47	
0.2 / 0.2				120Kts		130 /05	

Dublin to Belfast

Your routing for the leg to Belfast will take you to the West of Dublin before turning North. This is to keep you clear of the military operating areas to the North of Dublin and minimise the time you are over the sea. Although an engine failure on a simulator is usually just an opportunity to grab another coffee, we want to genuinely simulate the real thing and most pilots would choose to fly over land where the option exists.

Your initial route out from the airport to RANAR is straightforward enough. Route directly to the Golf Alpha Romeo and then gently drift onto the B1 centreline on the Dublin 293 radial until you reach 38 miles DME. Now comes the hard part. If you have a suitable map and are good at map reading, you may be able to confirm your route, but otherwise you will have to aim for the intercepts of VOR DME's. Cavan is 48 DME from DUB on the radial 315 and Monaghan is on the 232 radial from BEL on 117.2 at 36 DME. Again, if you cannot receive these beacons you will be stuck with dead reckoning, that is flying your plan exactly until you can positively determine your location. If you fly accurately and carry out regular FREDA checks, you should actually end up pretty close to your destination. You may be tempted to let technology fly the plane for you, however there is a danger with this because hand flying without using the autopilot can be the most accurate way to fly the smaller flight simulation planes. Once you have set a plane on a course and trimmed it well, it is not too difficult to fly very accurately by just making tiny adjustments and using tiny control inputs.

After the cross country legs to Portadown you approach the VRP which is 16 miles from Belfast on the 215 radial. When you call for zone entry clearance, you are told to report at Portadown without a clearance to progress any further. When you report at Portadown, you are told to hold there for 5 minutes and then route direct to the airport not above 1500 ft QNH, and report entering the zone. You must now orbit for 5 minutes. This does not have to be a standard holding procedure, as they are purely holding you there to keep you out of the way of some other traffic. Your task is to circle over Portadown making sure that you cross over the town for the last time after 5 minutes and head for the zone boundary. If you are flying accurately you will enter the zone 3.5 minutes later, so you call them as requested. As you enter the zone you are vectored North for five minutes and then told to report right base for runway 17. When you arrive on right base you are finally given clearance to land.

Whilst you often are given a smooth passage into an airport's control zone and then on to landing, it is not always guaranteed. It is quite common to be held and vectored in the way you have just been at Belfast. Airports can be very busy places and the controllers need to mix planes large and small, fast and slow, and to do it all safely with adequate separation. It would be unusual to be given frequent height changes, in the UK at least, unless it was part of your normal approach. You could have your decent delayed or brought forwards, but it is unlikely you would be told to climb back up again after you have been previously given clearance to descend.

✈ FLIGHT LOG — 6 Waypoints

| Position | ID/Freq | | ETA | RTA | ATA | TTLT | GDTG |
ALT/MSA	AWY	TRM	HDM	IAS	TIM	W/V	DST
Belfast Alder-grove	EGAA					0	145
2.0 / 2.0		358	359	120Kts	6	130/05	12
Ballymena	VRP407					6	133
	B1	039	041	120Kts	19	130/05	40
Machrihanish	MAC116					25	93
3.0 / 2.7		089	091	120 Kts	18	130/05	37
Prestwck	PIK355					43	56
3.0 / 2.0		076	078	120 Kts	21	130/05	44
West Linton	VRP					64	
3.0 / 2.9		365	007	120 Kts	6	130/05	12
Edinburgh	EGPH					64	
0.1 / 0.1				120 Kts		130/05	

Belfast to Edinburgh

So we have reached the final leg of this tour of some of the major cities in the British Isles. Before you take off, you tell the controller that you are "Tower Golf Bravo Sierra Mike Hotel ready to copy clearance" and their reply is "Golf Bravo Sierra Mike Hotel Tower cleared to leave the zone at Ballymena Special VFR not above 1500 on the QNH 1013". So you say this back to her exactly as she has just said it to you, but putting the word Tower first again, and then tell her that you are "Ready for departure". She immediately responds that you are "Clear takeoff Runway 17, climb straight ahead to 1500 feet before turning right onto track, surface wind 130 degrees 5 knots". All you have to do is to accept the clearance with "Straight ahead to 1500 feet before turning right, clear take off" and away you go.

Ballymena is another VOR DME fix from BEL and is 12 miles on radial 360. If you have managed to fly this far, you will have no problem at all finding it. Just at this moment your co-pilot, who has dutifully been manning the radios for you most of the time says that he has never seen the Giant's Causeway. It's not too far off your track, and despite your doubts about being able to spot if from the plane, you decide to fly over it. It should be easy to find, you just have to track up the 352 Belfast radial until you reach the coast. Once you are there, you find that you cannot see it (unless you have some really fancy scenery installed), so as you cross the coast at 37 DME, you turn to fly directly to Machrihanish.

If you have already completed the IFR tutorial you will have no trouble with the leg from Machrihanish to Prestwick and then backtracking the NDB there to West Linton, which can be located using the TLA on 113.8 on the 005 radial at 15 DME. This time, as you are given your clearance to enter the zone, you are told to turn heading 030 and maintain 3000 feet. Five minutes later you are instructed to report right hand downwind for 06. This is not Edinburgh's best runway for you in the current wind, but with all the heavy jet traffic, they are using the long runway for all traffic. Your downwind and base leg must be at or above 1000 feet above the airport, so either fly on the airport QFE or at 1200 feet on the QNH. You call "downwind" as you pass abeam the threshold of runway 24 and the controller tells you to carry out one orbit left. What you have to do here is to turn left through a complete circle, there is no specific angle required for this, so simply hold altitude and speed and roll into a 30 degree turn until you are back on your original downwind track. You are then "cleared to finals number two", and by the time you report finals the plane in front of you has obviously got out of your way because you are cleared to land.

Shutdown on the Apron

You taxi to the apron on the South side of the main runway and shutdown, so ending a perfect day of flying. No doubt you reached every waypoint exactly as planned and were never uncertain of position. You will have covered over 850 statute miles. It's quite likely that the sun is setting in a gin clear sky, after all, why add nasty weather to make things difficult for yourself on a big challenging flight like this. You look down at your flight logs and see how long each flight took you, and think to yourself, "That wasn't so bad after all, I reckon I could find my way just as easily in the worst weather nature can throw at me". Well that's the beauty of flight simulation; you can start again, turn on the weather, turn off the sunshine and see how good you really are!

COMPETITION

As this is a large flight, to get the most from it you must fly it accurately. Therefore we have decided to lay down the challenge to see how close **you** can get to the time we took in the Cessna 182S for Flight Simulator. Follow the flight log, fly the route exactly as described here and record the time you took on each of the individual sectors to the nearest second. The time you record should be your airborne time, or in other words, the time your wheels leave the runway on take off until they touch the ground again on landing. You get the flight from Jersey to Gatwick to warm up, so the times that will count are:

☒ Jersey to Gatwick (warm up leg)
☑ Gatwick to Cardiff
☑ Cardiff to Dublin
☑ Dublin to Belfast
☑ Belfast to Edinburgh

Remember that this is not the just time on the flight logs, as these do not account for IAS versus TAS, turns, climb, decent or any vectoring around the sky given by ATC.

We have not included the leg from Jersey to Gatwick in this competition in case Jersey is not featured in your simulator. Any reasonably comprehensive simulator covering the British Isles should include the five remaining airports.

When you have these four times visit TecPilot's Aviation Challenge Web Site at http://www.tecpilot.com and fill in our competition entry form. Don't forget to include your e-mail address and full name. Insert the minutes and seconds for each leg of the flight as you see them on the form. Please double check what you enter. Alternatively, enter the information on the back of a post card and send it to:

The **Flight** of your Life
TecPilot Publishing
76 Victoria Road
Parkstone, Poole
Dorset, BH12 3AF
United Kingdom

One winner

The person who gets closest to our time before the end of October, 2002 will receive a software prize to the value of £200.00 plus a Corporate Membership to the members only area of our web site worth £35.00.

If there is a draw, the winner will be decided by a tie breaker question, which is to complete the phrase "*Flight simulators are better than the real thing because* ____" in less than 30 words. Please include your tiebreaker answer if you are sending in your entry on a post card!

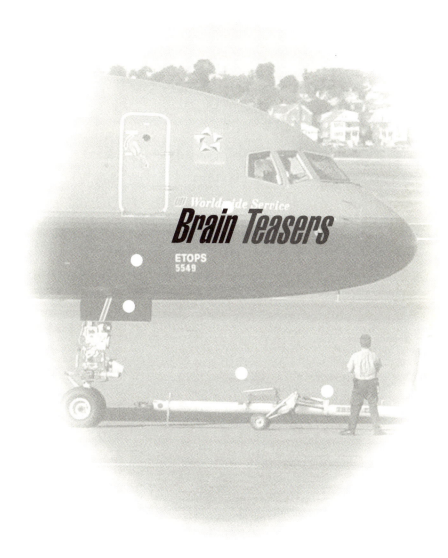

Brain Teasers

The 52 Week Aviation Quiz

For those of you who like the challenge of a long quiz, here is one question for every week of the year. The first section contains all the questions and the second section has a helpful hint for every question. If you don't know the answers, try to hunt them down. The answers are given at the back of this yearbook. Have fun!

Questions 1-13

Hints

#	Question	Hint
1	You want to fly to a VOR on the 056 radial, you tune the radio, ident it and set 056 on the OBS. Your heading is 225. The needle shows full fly left and the "TO" flag is showing. Which way do you turn and what is a sensible heading to fly initially?	Your heading has no effect on the indications on the VOR instrument.
2	What is an SRA?	It's a type of approach that is easy in small slow planes and difficult in big fast ones.
3	Name the two types of de-icing that can be found on most light aircraft.	One of them can be found on the FS2000 Cessna panels and the other can be found in a road vehicle.
4	You are lost in the UK. If you tune your NAV1 to 112.5 and the VOR needle centralises with the OBS on 095. You then retune to 117.5 and setting the OBS to 205 centralises the needle. Where are you and what would you expect the beacon Morse idents to be?	Fly high in a fast jet from one end of the country to the other.
5	You are flying cross country on your planned heading of 256 Magnetic and have split the current sector into four to check your navigation en-route. At the quarter way point you realise that you have been blown off course by 7 degrees to the right. What heading should you steer to put you back on track and what heading to then take you to the next waypoint?	You must both correct and compensate for the error, but not always at the same time.
6	You are flying from London to New York. You calculate the heading for the shortest distance as 288 degrees true and amazingly there is no wind over the whole Atlantic. If you hold this heading very accurately for 3000 nautical miles, will you end up in New York?	A circle is the shortest distance between two points on the world?
7	You are flying directly from London Heathrow to Manchester on a continuous heading of 335. Are you flying in a straight line?	Weston-Super-Mare to Liverpool is a straight line, and so is Libreville to Mbandaka.
8	The Air Speed Indicator (ASI) is painted with peripheral arcs in green, yellow and white. What are they for?	They are bound by VSO, VSI, VFE, VNO and VNE.
9	Why is the leading edge of some wings black?	It's covered with inflatable rubber, but why?
10	What is the difference between a radial and a rotary engine?	The rotary engine was used in the First World War, whereas radial engines are still used today.
11	What does LITAS stand for and how is it used?	Have a word with your PAPIS.
12	An ATC tells you to squawk 3662 with ident, what should you do.	It's all to do with Secondary Surveillance Radar.
13	On the Jeppesen approach plate for Filton, the ILS frequency for runway 27 is given as *110.55 IFB. Why is the asterisk there?	Time is the key.

The Good Flight Simmer's Guide

Questions 14-26

14	What is a rate one turn and what is it used for?	It's odd that it's called a rate one turn, when the numbers associated with it are 2 and 3 and 360.
15	What are categories A, B C and D on an approach plate?	You can work out which one to use by stalling an aeroplane.
16	Approach plates show some heights in brackets or in a lighter font, what are they?	What pressure would you set your altimeter when flying an approach and landing?
17	What's the difference between orbiting and a holding pattern?	Precision is the key.
18	You are about to take up a 120 knot hold heading based on heading 045 to an NDB and then turning left. The wind is 20 knots from the North, what headings and times would you expect to fly in the hold.	Work out your headwind and crosswind components and then apply the holding pattern rules of thumb.
19	Why do some light planes have wedges on the front of the wing near the fuselage?	They make that bit of the wing stall first.
20	What is wing warping?	This Wright way was shown to be the wrong way to do things.
21	Your gyro Direction Indicator has failed so you are instrument flying using the magnetic compass as the only indication of your heading. You are flying West and turning the shortest way onto South. You watch the compass for the correct heading indication before rolling the wings level, but what would the correct heading indication be?	Use the UNOS acronym and 30 degrees.
22	You are planning to land in a light plane just after a B747 has landed. How long would you wait after it has landed before you would be prepared to land, and why?	A Jumbo Jet has to disturb an awful lot of air to fly.
23	Why do pilots like to take off into wind?	Ground speed is irrelevant to
24	As you fly over an aerodrome, you see a big white "T" on the ground. What does it mean?	This is one of several ground signals that are used to communicate with non-radio aircraft
25	You have taxied onto the apron under the guidance of the marshal. He holds his light sticks straight above his head and moves them together from side to side above his head. What does he want you to do?	His next signal will probably involve an action that looks like he is cutting his own throat.
26	You take of from your favourite airport having set the altimeter to the aerodrome QNH. You are flying towards the centre of a depression across flat ground and keeping the indicated height on your altimeter constant. Will you get closer to the ground or further away?	Depressions usually have clouds, and when things get bad, they get worse.

The 52 Week Aviation Quiz

27	You are flying on heading 315 and the ADF shows a relative bearing of 235. What is the applicable QDM?	*QDM is the magnetic heading to the beacon without any correction for wind.*
28	What is a QGH procedure?	*You don't need any radio navigation instruments for this approach procedure, just the normal VHF communications radio.*
29	What has "Charley Charley" got to do with Morse Code?	*Pretty much the same as Q's "Here Comes the Bride".*
30	You receive a weather report that states the weather as CAVOK. What weather does this indicate and could it be overcast?	*CAVOK generally indicates excellent flying conditions, but....*
31	You are flying upside down on a Northerly heading. You want to stay inverted and turn the shortest way onto an Easterly heading. Which way would you initially move the stick and which rudder pedal would you push?	*A plane upside down is pretty much the same as a plane the right way up with you hanging upside down from the bottom. The control surfaces move the same way, just think about what you have to do with the controls to make it happen.*
32	You a flying a Reverse Cuban Eight and are flying upwards at 45 degrees but upside down. What reading would you see on the accelerometer?	*Sine and Cosine 45 are both about 0.7.*
33	You are flying in class G airspace at night and you see the red wing tip navigation light of another plane and its flashing anti-collision light. They appear to be at about the same height as you and are stationary in the sky to your right. What should you do?	*The red light is on the port or left wing, and worrying would be a good place to start.*
34	ou are about to level out having climbed to your cruising height. Of course you need to push the joystick forward, but you also need to trim the aircraft and set the throttle, mixture and propeller RPM. What order should you do these in?	*It would be sensible to level off before reducing power, otherwise you would loose speed rather quickly.*
35	You have been flying aerobatics in your aeroplane and have accidentally entered an upright spin to the left. You are several thousand feet high with a parachute strapped on, what should you do?	*It would be expensive to bail out too early, but of course foolhardy to leave it too late.*
36	Why do some planes have a big broad vertical white line painted on the instrument panel?	*It helps with spin recovery.*
37	The angle of instrument approach paths varies from aerodrome to aerodrome, but what is the most common angle used?	*It's somewhere between 2 and 10 degrees.*
38	What does the radio transmission of "Roger" mean?	*It's not the same as Wilco.*
39	The transition altitude at Cardiff is given as 4000 feet. What does this mean?	*There is also a transition level.*

Questions 40-52

40	You have just left the Teeside Control Zone and the air traffic controller tells you that the Tyne is 1032 and the Barnsley is 1033. What does this mean?	*The full description would normally include the term QNH.*
41	You have just taken off in your Cessna 182 and at 500 feet the engine fails. Should you turn back and land on the runway?	*Try it out on your flight simulator. Ideally with someone sneaking up on you and suddenly closing the throttle any time over the next few weeks, this way you will be unprepared for it.*
42	The flaps in your light plane have failed in the retracted position and you are on final approach to land. You realise that you are too high. What should you do?	*If you are too high, you could go around and try again, but there is another option that is regularly used by planes that have been*
43	What is the most common direction for circuits and why?	*Which side of the cockpit does the captain sit on?*
44	You are flying along in a twin-engine aircraft when the right engine fails. Which rudder pedal would you expect to press to keep the slip ball in the middle?	*The slip ball will have moved out to the left.*
45	You are flying just below the overcast cloud base, but the temperature is 10 degrees Celsius above freezing. Are you likely to be prone to carburettor icing?	*If you are just below the cloud base, the humidity will be high.*
46	What is QSY?	*You would hear it said over the radio, even though it is now no longer correct parlance.*
47	Why should you include the suction gauge in your instrument scan?	*Which instruments are powered by the suction?*
48	What do "toppling" and "caging" normally refer to in an aeroplane?	*They are usually associated with aerobatics.*
49	You are on final approach to land in your Cessna and realise that you are a bit too low, what should you do?	*If you simply pull back on the stick, you will slow down too much.*
50	How many gee would you pull in a properly balanced 60 degree banked level turn?	*It's the reciprocal of the cosine of the angle of bank*
51	What is the difference between the turn indicator found on older planes and the turn coordinator found on modern planes?	*The turn indicator only gives the rate of turn, the turn coordinator gives something else as well.*
52	What is the "beta range" on a propeller?	*You only find this on some variable pitch propellers. They spend most of their time in the alpha range.*

Aviation Crossword

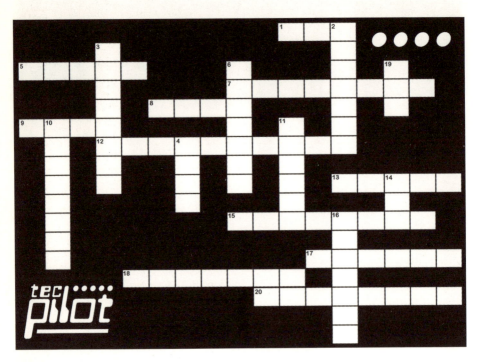

Across

1.	Abbreviation for an onboard computer
5.	Where controllers operate from at airports
7.	Manipulate this to increase thrust
8.	Sounds like the inside of an egg!
9.	Movements to and from the runway
12.	Course Plotting
13.	Keeps aircraft aloft at slow speeds
15.	Reverse on the ground with assistance
17.	Replenish engine stocks
18.	Does not float, it flies and looks like a cigar

Down

2.	Chief Pilot - the boss.
3.	Gathering place for passengers
4.	Type of aircraft fuel
6.	Who brings refreshments during a flight
10.	Another word for an aeroplane
11.	The flyable version of public transport
14.	Code given to an Airport
16.	First across the English Channel in a plane
19.	Abbreviation of a guided landing device

Aviation Word Search

Word list:

- Airbus
- Airport
- Altimeter
- Beechcraft
- Bleriot
- Boeing
- Cessna
- Engines
- Flaps
- Flight
- Fly
- FMC
- Glider
- Propeller
- Simulator
- Spoilers
- Takeoff
- TecPilot
- Thrust
- Traffic
- Undercarrage
- Wings

a-Z
Boxed Add-on Products
Adventures

TecPilot Ratings	
Forget It!	✦ ✦ ✦ ✦ ✦
Below Average	✦ ✦ ✦ ✦ ✦
Average	✦ ✦ ✦ ✦ ✦
Above Average	✦ ✦ ✦ ✦ ✦
Excellent!	✦ ✦ ✦ ✦ ✦

Title: Adventure 2000

Publisher/Developer: Apollo Software

Web Site: http://www.apollosoftware.com/

TecPilot Rating:

Compatible with: FS98

Requirements: Pentium 266 minimum (recommended), 64 Mb Ram, Microsoft Flight Simulator 98, CD Rom Drive, Soundblaster Compatible Sound Card, 2D/3D Card, Headphones or Speakers recommended.

Description: Adventure 2000 is a very simple navigation tool and flight planner. After your flight plan is entered, you simply fly the headings it instructs. Adventure 2000 will also export flight plans to First Class 747 if you have it.

In A Word: Simple, maybe too simple.

Title: Airline Flights 1

Publisher/Developer: Aerosoft

Web Site: http://www.aerosoft.com

TecPilot Rating:

Compatible with: FS98

Requirements: PC CD-ROM. Pentium 333 or higher. Windows 95/98. Microsoft Flight Simulator 98. 32mb RAM. SVGA graphics card. 2x CD-ROM or greater required

Description: Original voices of pilots and controllers.. Includes the flights Frankfurt-Salzburg, Paderborn-Amsterdam, Frankfurt-Madrid, Duesseldorf-Paris, Paderborn-Gran Canaria, Teneriffa-Duesseldorf, and loads more. Plus the aircraft Bae 146, Airbus A340, Airbus 330 and the Concorde.

In A Word: Lets hope you're multilingual

Title: Airline Flights 2000

Publisher/Developer: Aerosoft

Web Site: http://www.aerosoft.com

TecPilot Rating:

Compatible with: FS2000, FS98

Requirements: PC CD-ROM. IBM PC & compatibles. Pentium 200 or higher. Windows 95/98. Microsoft Flight Simulator 2000. 32mb RAM. SVGA graphics card required. 2x CD-ROM or greater required

Description: For these adventures the voices of professionals were used. The flight begins on the docking bridge, and all procedures beginning with engine start must be completed. As an extra bonus 5 aircraft are included (3 x B747, 2 x B767).

In A Word: Boring. Hard to understand sounds

Title: ## California Adventure Collection

Publisher/Developer: Aerosoft

Web Site: http://www.aerosoft.com

TecPilot Rating: ✦ ✦ ✦ ✧ ✧

Compatible with: FS98

Requirements: PC CD-ROM. IBM PC & compatibles. Pentium 200 or higher. Windows 95/98. Microsoft Flight Simulator 98. 32mb RAM. SVGA graphics card required. 2x CD-ROM or greater required

Description: 15 professional IFR-approaches and 2 high-quality adventures in California. In the adventures, you can choose to fly the approach before you start the whole adventure. All necessary approach charts are included.

In A Word: Interesting enough.

Title: ## Captain Speaking

Publisher/Developer: JustFlight

Web Site: http://www.justflight.com

TecPilot Rating: ✦ ✦ ✦ ✧ ✧

Compatible with: FS2000

Requirements: 350 Mhz Processor, 64Mb RAM, 290Mb HD space.

Description: Captain Speaking allows you to take 37 different airline flights, accompanied by authentic air traffic control. More the 22,000 voice recordings were included, In 3 different languages. The program even makes you change squawk and altitude changes when flying to a different continent. The program also equips you with a GPSW, with MAYDAY instructions.

In A Word: Takes ATC to a whole new flight level!

Title: ## FlightSim Commander

Publisher/Developer: Aerosoft

Web Site: http://www.aerosoft.com

TecPilot Rating: ✦ ✦ ✦ ✧

Compatible with: FS2000

Requirements: PC CD-ROM. IBM PC & compatibles. Pentium 200 or higher. Windows 95/98. Microsoft Flight Simulator 2000. 32mb RAM. SVGA graphics card required. 2x CD-ROM or greater required

Description: Flight Sim Commander is the all-in-one flight planner. Features include, great circle route planning, ATC with voice, automatic flight planning, auto-pilot take off and landings, moving maps, even a flight recorder!

In A Word: Navigation made simple

Title: **Lufthansa Pilots**

Publisher/Developer: Sky Design

Web Site: http://www.skydesign.de

TecPilot Rating: ✦✦✦✦✦

Compatible with: FS2000

Requirements: PC CD-ROM. IBM PC & compatibles. Pentium 200 or higher. Windows 95/98. Microsoft Flight Simulator 2000. 32mb RAM. SVGA graphics card required. 2x CD-ROM or greater required

Description: Real world pilots came together to create this fresh and original package for FS2000. Lufthansa Pilots puts you though realistic training missions, based of the real Lufthansa Airline syllabus. The package includes, planes, panels, scenery, and adventures.

In A Word: Nice, German precision at its best

Title: **ProFlight 98**

Publisher/Developer: AETI

Web Site: http://www.simpilot.com

TecPilot Rating: ✦✦✦✦✦

Compatible with: FS98

Requirements: CD-ROM Sound Card, Pentium 2, 3-D Accelerator Card, Win95/98

Description: A very popular Adventure Planner – Simply fill out a simple flight plan to create a digitized voice adventure between thousands of airports anywhere in the world.

In A Word: Very popular and intuitive

Title: **Pearl Harbour**

Publisher/Developer: Flight One

Web Site: http://www.flight1.com

TecPilot Rating: ✦✦✦✦✦

Compatible with: CFS2

Requirements: 600 Mhz Processor, 128Mb RAM, Sound card and 3D Accelerator card

Description: Flight One's Pearl Harbor allows you be a Curtis P40B pilot in one of the greatest attacks in US history, the attack by Japan on Pearl Harbor. All the missions are historically accurate. The Curtis P40B also includes custom sounds, realistic flight modeling, and a virtual cockpit.

In A Word: The ultimate answer to "what if".

Title: **ProFlight 2000**

Publisher/Developer: Vmax

Web Site: http://www.simpilot.com

TecPilot Rating: ✦ ✦ ✦ ✦ ✦

Compatible with: FS2000, FS98

Requirements: Pentium 400 (recommended), 64 Mb Ram, CD Rom Drive, Soundblaster Compatible Sound Card, 2D/3D Card, Headphones or Speakers recommended.

Description: AETI's FS2000 version of their popular ProFlight '98 is one of the most complete ATC and flight planning programs for flight simulator. There are over 40 complete voice sets, featuring speakers from around the world. Dynamic chatter is also included.

In A Word: Got to love over achievers!

Title: **WinPlanner Plus 4**

Publisher/Developer: Lago

Web Site: http://www.lagoonline.com

TecPilot Rating: ✦ ✦ ✦ ✦ ✦

Compatible with: FS98, CFS

Requirements: Pentium 266 minimum (recommended), 64 Mb Ram, CD Rom Drive. 2D/3D Card.

Description: WinPlanner is LAGO's one and only flight planner released. The program allows one to automatically, or manually create flight plans. The package not only allows you to export your flight plans to other popular navigational software but achieve many other tasks as well.

In A Word: Complete and compatible.

The Good Flight Simmer's Guide

a-Z
Boxed Add-on Products
Aircraft

AIRCRAFT

Title: 3 Great Planes (Blue Box)

Publisher/Developer: Abacus

Web Site: http://www.abacuspub.com

TecPilot Rating: ✧✧✧✧✧

Compatible with: FS2000

Requirements: PC CD-ROM. IBM PC & compatibles. Pentium 200 or higher. Windows 95/98. Microsoft Flight Simulator 2000. 32mb RAM. SVGA graphics card required. 2x CD-ROM or greater required.

Description: Package contains three popular private aircraft including the Cessna Caravan, DHC6 Twin Otter and Piper Cherokee.

In A Word: 3 of the best!

Title: 3 Great Planes (Gold Box)

Publisher/Developer: Abacus

Web Site: http://www.abacuspub.com

TecPilot Rating: ✧✧✧✧✧

Compatible with: CFS, FS2000

Requirements: PC CD-ROM. IBM PC & compatibles. Pentium 200 or higher. Windows 95/98. Microsoft Flight Simulator 2000. 32mb RAM. SVGA graphics card required. 2x CD-ROM or greater required.

Description: Package contains several popular aircraft including the F4U-5 Corsair, P51 Dakota Kid Mustang, Supermarine Spitfire.

In A Word: Good things come in small packages

Title: 767 Pilot in Command

Publisher/Developer: Wilco Publishing

Web Site: http://www.wilcopub.com

TecPilot Rating: ✧✧✧✧✧

Compatible with: FS2000

Requirements: Microsoft® Flight Simulator 2000 - Pentium 450 MHz - 128 Mb Ram — 3D graphics accelerator card.

Description: The 767 Pilot in Command package puts you in some of the most terrifying situations that an airline pilot could ever experience. You'll challenge fires, depressurisations, and other problems. The plane and panel included are both recreated in top quality.

In A Word: A classic in the making!

AIRCRAFT

Title: **737-500 for Fly!**

Publisher/Developer: Wilco Publishing

Web Site: http://www.wilcopub.com

TecPilot Rating: ✦ ✦ ✦ ✦ ✦

Compatible with: Fly!

Requirements: MAC version: FLY! or Fly!2K From TRI, FLY! v1.01.83 patch or later, Power Macintosh G3 233MHz or better with 32MB RAM, a 22 MB hard disk drive for a minimum install and a 40 MB drive for full install. MAC version: FLY! or Fly!2K From TRI, FLY! v1.01.83 patch or later, Power Macintosh G3 233MHz or better with 32MB RAM, a 22 MB hard disk drive for a minimum install and a 40 MB drive for full install. PC and MAC compatible.

Description: Wilco remains one of the few company's who ventured into the world of Fly! add-ons. The 737 package includes an ultra high resolution, detailed panel, as well as 14 different aircraft liveries. The exterior leave something to be desired, the panel raises an eyebrow.

In A Word: The Fly!er's friend

Title: **Airline Pilot 1**

Publisher/Developer: Aerosoft

Web Site: http://www.aerosoft.com

TecPilot Rating: ✦ ✦ ✦ ✦ ✦

Compatible with: FS2000, FS98

Requirements: Microsoft® Flight Simulator 2000, Pentium 450 MHz, 128 Mb Ram, 3D graphics accelerator card

Description: Airline Pilot 1 is quite possible the most complete collection of adventures for flight sim. The 6 flights all includes outstanding audio, as well as various charts. Several planes are also include in the package.

In A Word: Not that hot!

Title: **Airforce**

Publisher/Developer: Flight and Cockpit

Web Site: http://www.pcflight.ch/

TecPilot Rating: ✦ ✦ ✦ ✦ ✦

Compatible with: FS2000

Requirements: Microsoft FS2000 Professional Version, AMD or Pentium higher 500 MHz, CDROM Drive, 64Mb Ram

Description: Fly the Hunter J-4083, including panel from Swiss Squadron base 21. Included are dynamic flights. This Package includes eight Swiss Military Airports including Alpnach, Ambri, Interlaken, Mollis, Meiringen, Raron, Stans and Ulrichen.

In A Word: Whoosh! Where's my gloves!

AIRCRAFT

Title: A319, A320, A321, 2nd Generation

Publisher/Developer: Flight and Cockpit

Web Site: http://www.pcflight.ch/

TecPilot Rating: ✦ ✦ ✦ ✦

Compatible with: FS98

Requirements: Pentium 266 minimum (recommended), 64 Mb Ram, Microsoft Flight Simulator 98, CD Rom Drive. 2D/3D Card.

Description: The package includes the A319, A320, A321 for FS98. Each plane includes an exterior model as well as new flight dynamics. Each plane also includes a panel in 1024x786 resolution, that includes glass cockpit instruments.

In A Word: In its time, The Next generation

Title: Air Power Cold War (1947-1970)

Publisher/Developer: JustFlight

Web Site: http://www.justflight.com

TecPilot Rating: ✦ ✦ ✦ ✦

Compatible with: FS2000, FS98, CFS

Requirements: Microsoft Flight Simulator 2000/98 or Combat Flight Simulator, Pentium II 266 PC, Windows 95, 98, 32 Mb Ram, 398Mb Hard drive space for FS2000 Install, 173Mb Hard drive space for FS98 and CFS install, CD ROM Drive, Sound Card.

Description: Fly the might of NATO and Warsaw Pact Air Forces in this Incredible Expansion. The mighty air forces of NATO and the Warsaw Pact clash again... This amazing expansion provides 29 legendary aircraft, together with scenery and missions* for Flight Simulator 2000/98 and Combat Flight Simulator!

In A Word: Come out of the cold and into the power zone!

Title: Airbus 2000 Special Edition

Publisher/Developer: JustFlight

Web Site: http://www.justflight.com

TecPilot Rating: ✦ ✦ ✦ ✦

Compatible with: FS2000, FS98, CFS

Requirements: Pentium PII 266 Mhz or higher, 3D graphics card, 64 Mb Ram

Description: The Airbus aircraft have been re-created in JustFlight's Airbus 2000 package. More than 50 different planes are included, all featuring a GPWS, new age exterior designs, and a unique panel for each plane, which even includes a photo-realistic jump seat views.

In A Word: They boldly went where no designer has gone before.

Title: ## B-747-357 Reality Package

Publisher/Developer: Flight and Cockpit

Web Site: http://www.pcflight.ch/

TecPilot Rating: ✦ ✦ ✦ ✦ ✦

Compatible with: FS2000

Requirements: Microsoft FS2K or FS98 - Pentium higher 300 MHz -CD-ROM

Description: Includes artificial horizon, speed and altitude indicators, radio altimeter, vertical speed indicator, horizontal situation indicator (HSI). Autopilot with flight director, auto throttle and systems for full automatic landings.

In A Word: That's what I call complete!

Title: ## Boeing 757/767 For Fly!

Publisher/Developer: Aerosoft

Web Site: http://www.aerosoft.com

TecPilot Rating: ✦ ✦ ✦ ✦ ✦

Compatible with: Fly!, Fly!2k

Requirements: Pentium 266 minimum (recommended), 64 Mb Ram, Microsoft Flight Simulator 98, CD Rom Drive. 2D/3D Card

Description: The package includes 12 757s and 767s, painted in various airline schemes. The planes have been included with some of the most accurate flight dynamics, and panels ever for Fly!.

In A Word: As close to the real thing as most of us will get!

Title: ## Boeing 777-200 Professional

Publisher/Developer: JustFlight

Web Site: http://www.justflight.com

TecPilot Rating: ✦ ✦ ✦ ✦ ✦

Compatible with: FS2000

Requirements: Pentium 266 minimum (recommended), 64 Mb Ram, Microsoft Flight Simulator 98, CD Rom Drive. 2D/3D Card.

Description: 777-200 is quite possible the most accurately designed aircraft ever. The body is extremely detailed, as well is the advanced panel. The sounds for 777 PRO are all digitally recorded.

In A Word: The only price for quality is system performance.

Title: **Combat Aces**

Publisher/Developer:	JustFlight
Web Site:	http://www.justflight.com
TecPilot Rating:	✦ ✦ ✦ ✦ ✧
Compatible with:	CFS2
Requirements:	Pentium III 450 or higher · 128Mb or more RAM · 16Mb or more 3D graphics card · Monitor capable of 1024 x 768 resolution · 3D sound card
Description:	Includes 20 aircraft, WWI scenery and 70 missions spread over six campaigns. 20 authentic aircraft, panels, scenery.
In A Word:	Great, if you remember and like this sort of thing. Fairly realistic.

Title: **Corporate Pilot**

Publisher/Developer:	Abacus
Web Site:	http://www.abacuspub.com
TecPilot Rating:	✦ ✦ ✦ ✧ ✦
Compatible with:	FS2000
Requirements:	Microsoft Flight Simulator 2000/98 Pentium II 300 PC, Windows 98, 98, 64 Mb Ram, CD ROM Drive, Sound Card.
Description:	Corporate Pilot painstakingly re-creates 5 of the most popular sleek and sexy biz jets in the industry. All include detailed panels, flight models, and sounds.
In A Word:	Ever thought you would call a plane 'sexy'?

Title: **Combat Pilot No.1 (Attack) Squadron**

Publisher/Developer:	JustFlight
Web Site:	http://www.justflight.com
TecPilot Rating:	✦ ✦ ✦ ✧ ✧
Compatible with:	CFS
Requirements:	Pentium PC. Up to 150Mb hard disk space for installation of entire collection. 17 inch monitor for maximum panel detail.
Description:	CP No. 1 includes seven carefully designed planes for CFS. The dynamics of these aircraft have been tested and approved by real pilots. Each plane's exterior features a round body, as well as moving parts.
In A Word:	A lot of energy goes a long way

AIRCRAFT

Title: Combat Squadron

Publisher/Developer: Abacus

Web Site: Http://www.abacuspub.com

TecPilot Rating: ✦ ✦ ✦ ✦

Compatible with: CFS2, CFS1, FS2000

Requirements: Pentium II 350 MHz CPU PC or greater. 17 inch or larger monitor for maximum panel detail. A good quality 3D Video Accelerator card capable of 1024x768 resolution with MB of RAM or more. Good quality Sound card.

Description: Combat Squadron includes 18 planes, designed by world renown designer, Terry Hill. Each plane includes sounds, panels, flight dynamics, as well as a exterior model. The package included-s some planes that are included in the stock CFS program

In A Word: Some are gems, some are rocks,

Title: Electra!

Publisher/Developer: JustFlight

Web Site: http://www.justflight.com

TecPilot Rating: ✦ ✦ ✦ ✦

Compatible with: FS98

Requirements: Windows 95/98 or NT; Pentium I Class 150 Mhz Processor; 12x CD Rom.Up to 35 Mb of total drive space.

Description: The VIP Group and The Cielo Company bring you this 2-CD multimedia experience! You will find it is a unique, one-of-a-kind tribute to the legendary Lockheed L-188 Electra airliner and highlights the use of that aircraft in its historic last passenger flying role up in Alaska and the Aleutian Islands.

In A Word: Takes me back to the day....

Title: Executive Jets

Publisher/Developer: Just Flight

Web Site: http://www.justflight.com

TecPilot Rating: ✦ ✦ ✦ ✦

Compatible with: FS2000

Requirements: P450Mhz, 64Mb RAM, CDROM, Sound Card, 3D Graphics Card

Description: This collection features some of the more expensive jets that are used today! The luxury planes feature a programmable FMC, and highly detailed exteriors, as well as interiors.

In A Word: Feel like a million bucks!

Title: **Eurowings Collection**

Publisher/Developer: Aerosoft

Web Site: http://www.aerosoft.com

TecPilot Rating: ✦✦✦✧✧

Compatible with: FS98

Requirements: Pentium 200, 64 Mb Ram, Microsoft Flight Simulator 98, CD Rom Drive. 2D/3D Card.

Description: The Eurowings Collection covers the airports of Paderborn, Dortmund and Nuremberg in addition to the entire Eurowings-Fleet of present and past. Airbus A319- Atr 72- Atr 42- BAe146 (Whisperjet)- Metroliner. An individually programmed panel is delivered with every aircraft.

In A Word: Where next boss!

Title: **Flight Deck (Blue Angels Edition)**

Publisher/Developer: Abacus

Web Site: http://www.abacuspub.com

TecPilot Rating: ✦✦✦✧✧

Compatible with: CFS, FS98, FS95

Requirements: Pentium 266 minimum (recommended), 32 Mb Ram, Microsoft Flight Simulator 98/95, CD Rom Drive. 2D/3D Card.

Description: Want to be a member of the US Navy's most elite? Flight Deck gives you the chance. By simulating carrier ops, you'll become familiar with how flying on and off a floating runway really feels! The package includes planes, and scenery.

In A Word: Right on the ball!

Title: **Flight Deck II**

Publisher/Developer: Abacus

Web Site: http://www.abacuspub.com

TecPilot Rating: ✦✦✦✧✦

Compatible with: CFS2, FS2000, FS2002

Requirements: Pentium 350 (recommended), 64 Mb Ram, CD Rom Drive. 3D Graphics Card.

Description: Update to the original Flight Deck Blue Angels product. Superb aircraft and scenery modelling together with many effects that are still "top gun" today.

In A Word: Much better this time around!

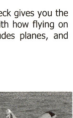

AIRCRAFT

Title: Fly The Best

Publisher/Developer:	Abacus
Web Site:	http://www.abacuspub.com
TecPilot Rating:	✦ ✦ ✧ ✧ ✧
Compatible with:	FS98, FS2002
Requirements:	Pentium 266 minimum (recommended), 64 Mb Ram, Microsoft Flight Simulator 98, CD Rom Drive. 2D/3D Card
Description:	Fly The Best, comes with 18 modern airliners. They all have detailed flight, and visual models, and instrument panels, all in different airline paint schemes. The bonus software of Instant Airplane maker, allows you to create your own paint schemes and more!
In A Word:	Decisions, Decisions!

Title: First Class 747

Publisher/Developer:	Apollo Software
Web Site:	http://www.apollosoftware.com/
TecPilot Rating:	✦ ✦ ✦ ✧ ✧
Compatible with:	FS2000
Requirements:	Pentium 266 minimum (recommended), 64 Mb Ram, Microsoft Flight Simulator 2000, CD Rom Drive. 2D/3D Card.
Description:	The First Class 747 is a unique product that focus on two accepts of the 747, the FMS, and EFIS. The products are designed to perform exactly like the real 747's navigation computers. Three FS98 747s are also included
In A Word:	First Class Quality!

Title: Fly to Hawaii: DC10

Publisher/Developer:	FlightSoft
Web Site:	http://www.flightsoft.com
TecPilot Rating:	✦ ✦ ✦ ✧ ✧
Compatible with:	FS2000, FS2002
Requirements:	Microsoft Flight Simulator 2000/98 Pentium II 300 PC, Windows 98, 64 Mb Ram, CD ROM Drive, Sound Card.
Description:	The Hawaii Airline DC10 was meticulously recreated by FlightSoft for the Fly to Hawaii add-on. The plane's biggest feature is it's round "wide body" fuselage. New techniques were used to give the plane an almost round look.
In A Word:	Mega Hit!

AIRCRAFT

Title: **Greatest Airliners: 737-400**

Publisher/Developer: Flight1

Web Site: http://www.flight1.com

TecPilot Rating: ✦ ✦ ✦ ✦ ✦

Compatible with: FS2000

Requirements: 600MHz Processor, 128Mb RAM, Sound Card, 3D Accelerator Card

Description: The Greatest Airliners: 737-400 was perhaps the most anticipated flight sim release in memory. The package includes the most detailed panel ever. It included the main forward panel, as well as overhead, and FMC views. The entire panel was made from actual cockpit photos. The airplane its self comes in several different liveries, and includes a program that allows you to create your own. The 737 features many complex components of the real 737, such as the Flap Load Relief system, and the flight announcement system.

In A Word: Not just one of the greatest airliners, one of the greatest add-ons!

Title: **Guardians of the Sky**

Publisher/Developer: Just Flight

Web Site: http://www.justflight.com

TecPilot Rating: ✦ ✦ ✦ ✦ ✦

Compatible with: FS2000

Requirements: P450Mhz, 64Mb RAM, CDROM, Sound Card, 3D Graphics Card

Description: In this "FS Classics" package, you get the detailed panels, and plane models of three sophisticated surveillance aircraft, including the Boeing E-3D Sentry.

In A Word: You don't need million dollar equipment to see these are beauties!

Title: **Harrier Jump Jet**

Publisher/Developer: JustFlight

Web Site: http://www.justflight.com

TecPilot Rating: ✦ ✦ ✦ ✦

Compatible with: FS2000, CFS, CFS2

Requirements: Pentium III 450 or higher · 64 Mb or more RAM · 3D Graphics accelerator card · Monitor capable of 1024 x 768 resolution · 3D Sound card

Description: The military's most remarkable aircraft is now available to fly in your favourite simulators! This stunning and comprehensive expansion provides the most accurate and detailed model ever seen of the legendary 'Jump Jet'.

In A Word: Vector that thrust. Then some more!

Title: **Jumbo Jet 2**

Publisher/Developer:	Data Becker
Web Site:	http://www.databecker.com
TecPilot Rating:	✦ ✦ ✧ ✧ ✧
Compatible with:	FS98, FS2000
Requirements:	PC CD-ROM. IBM PC & compatibles. Pentium 200 or higher. Windows 95/98. Microsoft Flight Simulator 2000. 32mb RAM. SVGA graphics card required. 2x CD-ROM or greater required
Description:	Takeoff and land while admiring the true-to-life scenery and terrain. Shut down two of the four engines and see what it's like to land with two. Negotiate the approach and choose from 100 airports from all around the world including Delaware, Maryland, Washington DC, Andrews Air Force Base, and Geneva.
In A Word:	Past it but available

Title: **Jumbo Jet 2000 (Version 3)**

Publisher/Developer:	Data Becker
Web Site:	http://www.databecker.com
TecPilot Rating:	✦ ✦ ✦ ✧ ✧
Compatible with:	FS2000
Requirements:	PC CD-ROM. IBM PC & compatibles. Pentium 200 or higher. Windows 95/98. Microsoft Flight Simulator 2000. 32mb RAM. SVGA graphics card required. 2x CD-ROM or greater required
Description:	Data Becker's Jumbo Jet 3.0 features 26 FS2000 compatible aircraft from the 747 family. Each airplane has it's own cockpit panel, with click-able buttons and knobs. The package includes a history on each aircraft, but a stop watch for mastering holding pattern.
In A Word:	Behind the times

Title: **Korean Combat Pilot**

Publisher/Developer:	Just Flight
Web Site:	http://www.justflight.com
TecPilot Rating:	✦ ✦ ✧ ✧ ✧
Compatible with:	FS2000, FS98
Requirements:	Pentium III 450 or higher, 64 Mb or more RAM, 3D Graphics accelerator card, Monitor capable of 1024 x 768 resolution, 3D Sound card
Description:	The Korean War - when the peak of piston-engined technology fought against fledgling jet fighters and the first all-jet duels took place. Now you can pilot the aircraft and fight the battles from this memorable period in flying history, when the eyes of the world were focused on the skies over Korea and the Sea of Japan.
In A Word:	Props v Jets—No match? Think again!

AIRCRAFT

Title:	**LTU 2001**

Publisher/Developer:	Aerosoft
Web Site:	http://www.aerosoft.com
TecPilot Rating:	✦ ✦ ✦ ✦ ✦
Compatible with:	FS2000
Requirements:	Microsoft Flight Simulator 2000/98 Pentium II 300 PC, Windows 98, 98, 64 Mb Ram, CD ROM Drive, Sound Card.
Description:	LTU from AeroSoft simulates the airline's entire fleet, including their new Airbus planes. Each plane is very detailed, and includes a high-end cockpit. Two of LTU's hubs (Dusseldorf and Palma de Mallorca) are also included in the package.
In A Word:	A whole airline on your computer!

Title:	**L-1011 Tristar**

Publisher/Developer:	JustFlight
Web Site:	http://www.justflight.com
TecPilot Rating:	✦ ✦ ✦ ✦ ✦
Compatible with:	FS2000. FS98
Requirements:	Pentium 166 minimum (recommended), 64 Mb Ram, Microsoft Flight Simulator 98, CD Rom Drive, SoundBlaster Compatible Sound Card, 2D/3D Card, Headphones or Speakers recommended.
Description:	This Tribute to the L-1011 includes a detailed flight model with panel in several different paint schemes.
In A Word:	The ultimate in historic airliners.

Title:	**Legendary Aircraft Collection 2000**
Publisher/Developer:	JustFlight
Web Site:	http://www.justflight.com
TecPilot Rating:	✦ ✦ ✦ ✦
Compatible with:	FS2000
Requirements:	This product is optimised for Flight Simulator 2000. Microsoft Flight Simulator 2000 OR Flight Simulator 98 OR Microsoft Combat Flight Simulator. Pentium II 266 PC. Windows 95/98 or NT (SP-3). 64 Mb RAM. 200 Mb of free hard disk space (to install entire collection). CD-ROM Drive.
Description:	This collection puts you in the cockpit o 226 plane from all points in history. From the Concorde to the Cub, these planes include detailed exteriors, panels, and sounds.
In A Word:	Every plane you'll ever need!

AIRCRAFT

Title: Luftwaffe Collection

Publisher/Developer: JustFlight

Web Site: http://www.justflight.com

TecPilot Rating: ✦ ✦ ✦ ✦ ✦

Compatible with: CFS, FS98, FS2000

Requirements: PC CD-ROM. IBM PC & compatibles. Pentium 300 or higher. Windows 95/98. Microsoft Flight Simulator 2000. 32mb RAM. SVGA graphics card required. 2x CD-ROM or greater required

Description: The Luftwaffe Collection is an extensive collection of Germany's military aircraft. Each plane is extremely detailed on the exterior, and includes a high-resolution panel. More than a dozen aircraft are included in the package.

In A Word: A little bit of everything

Title: Mad Dog 2000

Publisher/Developer: Lago

Web Site: http://www.lagoonline.com

TecPilot Rating: ✦ ✦ ✦ ✦ ✦

Compatible with: FS2000, FS98

Requirements: Microsoft Flight Simulator 2000, 98 or Combat Flight Simulator, Pentium II 266 PC, Windows 95, 98, 32 Mb Ram, 398Mb Hard drive space for FS2000 Install, 173Mb Hard drive space for FS98 and CFS install, CD ROM Drive, Sound Card.

Description: Mad Dog is LAGO's high quality rendering of the MD-80. The popular package includes both a simple, and complex panels, as well as one of the most detailed exterior flight models for Flight Simulator ever!

In A Word: Wow, no Woof!

Title: MD-11

Publisher/Developer: Flight and Cockpit

Web Site: http://www.pcflight.ch/

TecPilot Rating: ✦ ✦ ✦ ✦ ✦

Compatible with: FS98

Requirements: Pentium 266 minimum (recommended), 64 Mb Ram, Microsoft Flight Simulator 98, CD Rom Drive. 2D/3D Card.

Description: The package includes 36 different MD-11 liveries. The plane model has accurate stall and rotation speeds. The MD-11 panel is included as well. Developed by SwissAir pilots, the package is available in both FS98 and FS2000 formats.

In A Word: Great plane, great package.

AIRCRAFT

Title: Pacific Theatre Aircraft

Publisher/Developer: Abacus

Web Site: http://www.abacuspub.com

TecPilot Rating: ✦ ✦ ✧ ✧ ✧

Compatible with: CFS

Requirements: Pentium 266 minimum (recommended), 32 Mb Ram, Microsoft Combat Flight Simulator, CD Rom Drive. 2D/3D Card.

Description: Pacific Theatre puts you right in the middle of WW2's hottest battles. The package also puts some new scenery into CFS, such as aircraft carriers, as well as photo-realistic elevated mesh scenery covering many of the Pacific Islands.

In A Word: Beautiful scenery, horrible war!

Title: Pacific Combat Pilot

Publisher/Developer: JustFlight

Web Site: http://www.justflight.com

TecPilot Rating: ✦ ✦ ✦ ✧ ✧

Compatible with: CFS

Requirements: P200 MHz or higher CPU. 32MB RAM. 3D graphics accelerator card.

Description: Pacific Combat Pilot is a dense package which includes 27 new planes, two new campaigns, and 20 new mission. The package also puts some new scenery into CFS, such as aircraft carriers, as well as photo-realistic scenery covering many of the Pacific Island.

In A Word: Superior design for CFS!

Title: PBY-5 Catalina

Publisher/Developer: Abacus

Web Site: http://www.abacuspub.com

TecPilot Rating: ✦ ✦ ✧ ✧ ✧

Compatible with: CFS2, CFS1, FS2000

Requirements: Pentium II 350 MHz CPU PC or greater. 17 inch or larger monitor for maximum panel detail. A good quality 3D Video Accelerator card capable of 1024x768 resolution with MB of RAM or more. Good quality Sound card.

Description: The PBY-5 Catalina is a historic aircraft that's been used on both land, and sea, in civilian and military cases. The amphibious package includes an impressive exterior, along with a detailed panel. The panel even includes moving wipers!

In A Word: Time to get your feet wet.

Title: ## Phoenix 757-200

Publisher/Developer: JustFlight

Web Site: http://www.justflight.com

TecPilot Rating: ✦ ✦ ✦ ✦ ✦

Compatible with: FS2000

Requirements: Pentium III 700 MHz PC, 256 Mb RAM, 32 Mb 3D video accelerator card.

Description: 7 different airline liveries. Multiple cockpit views. Programmable FMC. Animated gear and control surfaces. 3D gear bays and rotating engine fans. Take the coveted left-hand seat in one of the world's most popular passenger airliners. This fantastic recreation for Flight Simulator 2000 is bursting with features and life-like detail.

In A Word: A Benchmark IFR aircraft. Extremely popular add-on when on launched.

Title: ## Phoenix Bonanza

Publisher/Developer: Just Flight

Web Site: http://www.justflight.com

TecPilot Rating: ✦ ✦ ✦ ✦ ✦

Compatible with: FS2000

Requirements: Pentium III 700 MHz PC, 256 Mb RAM, 32 Mb 3D video accelerator card.

Description: Three popular light aircraft are yours to fly in Flight Simulator 2000... Beechcraft V35 V Tail, 33C Aerobat and A36 Turboprop. Panoramic 360-degree cockpit views. Accurate panels. Animated gear and control surfaces. 3D gear bays and rotating props. Working lighting effects.

In A Word: Cheap and cheerful. Will keep fans of the Bonanza happy.

Title: ## Private Pilot

Publisher/Developer: Abacus

Web Site: http://www.abacuspub.com

TecPilot Rating: ✦ ✦ ✦ ✦ ✦

Compatible with: FS2000

Requirements: PC CD-ROM. IBM PC & compatibles. Pentium 200 or higher. Windows 95/98. Microsoft Flight Simulator 2000. 32mb RAM. SVGA graphics card required. 2x CD-ROM or greater required

Description: Private Pilot is a collection of 13 private propeller aircraft. All 13 planes feature moving parts (propellers, flaps, gear, etc), as well as detailed panels, as well as flight and visual models. The program also comes with aircraft checklists, and other great additions.

In A Word: Back to the basics!

AIRCRAFT

Private Wings

Publisher/Developer:	Data Becker
Web Site:	http://www.databecker.com
TecPilot Rating:	✦ ✦ ✦ ✦ ✦
Compatible with:	FS2000
Requirements:	PC CD-ROM. IBM PC & compatibles. Pentium 200 or higher. Windows 95/98. Microsoft Flight Simulator 2000. 32mb RAM. SVGA graphics card required. 2x CD-ROM or greater required
Description:	Private Wings features 36 quick and agile aircraft, each with a fully-functional cockpit and moving parts. Your first class flight is complete with original instrumentation, checklists, flight data, and radio with standby frequency. Clocks with stopwatch functionality allow for holding patterns and you can even log the number of hours in flight.
In A Word:	That's what I call complete!

Royal Air Force 2000

Publisher/Developer:	JustFlight
Web Site:	http://www.justflight.com
TecPilot Rating:	✦ ✦ ✦ ✦
Compatible with:	CFS, CFS2, FS2000
Requirements:	Pentium II 350 MHz CPU PC or greater. 17 inch or larger monitor for maximum panel detail. A good quality 3D Video Accelerator card capable of 1024x768 resolution with MB of RAM or more. 64 MB of RAM or more. Good quality Sound card.
Description:	The 25 aircraft in this package from the UK's RAF, all include very detailed exteriors, panels, and digitally recorded sounds!
In A Word:	**R**eally **A**bsolutely **F**antastic.

Stuka Dive Bomber

Publisher/Developer:	Data Becker
Web Site:	http://www.databecker.com
TecPilot Rating:	✦ ✦ ✦ ✦ ✦
Compatible with:	CFS
Requirements:	Pentium 266 minimum (recommended), 64 Mb Ram, Microsoft Combat Flight Simulator, CD Rom Drive. 2D/3D Card.
Description:	Choose from 10 historically exact dive bomber planes including Shark, Night Hunter, Ski, and Panzerknacker and brace yourself for an aerial combat adventure that puts you in the middle of a WWII dogfight. Takeoff and land while admiring the authentic scenery of Berlin in the 1940's.
In A Word:	Brace for impact!

Title: **Swiss Military 2000**

Publisher/Developer: Flylogic

Web Site: http://www.flylogicsoftware.com

TecPilot Rating: ✦ ✦ ✦ ✦

Compatible with: FS98, CFS

Requirements: Pentium 266 minimum (recommended), 64 Mb Ram, Microsoft Combat Flight Simulator, CD Rom Drive. 2D/3D Card.

Description: Features some of the top planes from the Swiss Airforce.

In A Word: Military in the mountains, it's gotta be good!

Title: **The Classic 747**

Publisher/Developer: AETI

Web Site: http://www.simpilot.com

TecPilot Rating: ✦ ✦ ✦ ✦

Compatible with: FS2000

Requirements: Pentium 300 minimum (recommended), 64 Mb Ram, Microsoft Flight Simulator 98, CD Rom Drive, Soundblaster Compatible Sound Card, 2D/3D Card, Headphones or Speakers recommended.

Description: The popular 747 was professionally recreated by AETI for the Classic 747 plane and panel package. The high resolution photo-realistic panel features navigation, overhead, and jump seat screens as well as the traditional pilots seat.

In A Word: One of the best!

Title: **The Next Generation**

Publisher/Developer: Vmax

Web Site: http://www.simpilot.com

TecPilot Rating: ✦ ✦ ✦ ✦

Compatible with: FS98

Requirements: Pentium 300 minimum (recommended), 64 Mb Ram, Microsoft Flight Simulator 98, CD Rom Drive, Soundblaster Compatible Sound Card, 2D/3D Card, Headphones or Speakers recommended.

Description: The planes of AETI's Professional Flight Collection are all packaged with highly detailed visual and dynamic models, as well as state of the art panels. The panels feature realistic avionics, and instruments. The package includes five planes in 24 liveries.

In A Word: A Skilfully Designed Package

AIRCRAFT

Title: **Tribute to the DC-10**

Publisher/Developer: Flight and Cockpit

Web Site: http://www.pcflight.ch/

TecPilot Rating: ✦ ✦ ✦ ✦ ✦

Compatible with: FS2000

Requirements: Microsoft Flight Simulator 2000/98 Pentium II 300 PC, Windows 98, 98, 64 Mb Ram, CD ROM Drive, Sound Card.

Description: This CD is a small tribute to the aging airliner. It includes a detailed exterior model as well as a panel and sounds. Technically replaced by Fly to Hawaii.

In A Word: A little small for a "tribute"

Title: **Tuskegee Fighters**

Publisher/Developer: Abacus

Web Site: http://www.abacuspub.com

TecPilot Rating: ✦ ✦ ✦ ✦ ✦

Compatible with: FS98, CFS

Requirements: Pentium 266 minimum (recommended), 64 Mb Ram, Microsoft Flight Simulator 98, Combat Flight Simulator, CD Rom Drive. 2D/3D Card.

Description: Tuskegee Fighter combines the brilliantly designed aircraft of Scott Nix, Alain L'Homme and Terry hill, with the scenery expertise of Tim Dickens, and tosses in realistic missions by Chris Steele. The package was created to honour the African-American airmen.

In A Word: A Skilfully Designed Package

Title: **VIP Classic Wings**

Publisher/Developer: JustFlight

Web Site: http://www.justflight.com

TecPilot Rating: ✦ ✦ ✦ ✦ ✦

Compatible with: FS2000

Requirements: Up to 150 Mb free hard drive space if you wish to install the full collection to the PC. (Not required for operation). 17 inch monitor.

Description: Ultimate Classic adds several new planes to your collection from the golden-age of aviation. All include planes, panels, and sounds.

In A Word: Those were the days!

AIRCRAFT

Title:

VIP Classic Airliners 2000

Publisher/Developer: JustFlight

Web Site: http://www.justflight.com

TecPilot Rating: ✦ ✦ ✦ ✦ ✦

Compatible with: FS2000

Requirements: Pentium II 266 PC. Windows 95/98 or NT (SP-3). 64 Mb RAM. 200 Mb of
                              free hard disk space (to install entire collection). CD-ROM Drive

Description: 16 historic airliners packaged with detailed exteriors, as well as rustic panels
                              for the VIP Classic Airliners collection. More then 200 variants, and liveries
                              are included in total. The planes also include there own handling notes.

In A Word: Master them all!

Title:

Wings Over China

Publisher/Developer: Abacus

Web Site: http://www.abacuspub.com

TecPilot Rating: ✦ ✦ ✦ ✦ ✦

Compatible with: CFS, FS2000

Requirements: PC CD-ROM. IBM PC & compatibles. Pentium 200 or higher. Windows 95/98.
                              Microsoft Flight Simulator 2000. 32mb RAM. SVGA graphics card required.
                              2x CD-ROM or greater required

Description: Wings over China is yet another one of Abacus's plane, scenery, and mission
                              packages for CFS. This time, the scenery covers China, as well as other
                              sectors of Southeast Asia. The planes included are a mixed bag of fighters,
                              and bombers.

In A Word: Explosively awesome

SCENERY

a-Z
Boxed Add-on Products
Scenery

The Good Flight Simmer's Guide

133

SCENERY

Title: ## African Safari Scenery

Publisher/Developer: Abacus

Web Site: http:/www.abacuspub.com

TecPilot Rating: ✦ ✦ ✦ ✦ ✦

Compatible with: FS98, FS95

Requirements: Pentium 266 minimum (recommended), 64 Mb Ram, Microsoft Flight Simulator 98, CD Rom Drive. 2D/3D Card.

Description: Visit famous national parks: Serengeti, Amboseli and Ngorongoro-Krater. You'll witness wildlife in natural settings-lions, elephants, zebras, wildebeast and more. Includes Mt. Kilamanjaro and the stately splendor of the Giza pyramids, major highways, waterways and mountain ranges of Africa and visit the most important cities.

In A Word: Grrrr!

Title: ## Airfield

Publisher/Developer: Papa Tango

Web Site: http;//www.papatango.com

TecPilot Rating: ✦ ✦ ✦ ✦

Compatible with: FS98

Requirements: Pentium 266 minimum (recommended), 64 Mb Ram, Microsoft Flight Simulator 98, CD Rom Drive, Soundblaster Compatible Sound Card, 2D/3D Card, Headphones or Speakers recommended, 160 Mb Disk Space.

Description: PapaTango recreates 300 dirt and grass strips throughout Europe in high quality in their Airfield add-on. Five aircraft are included in this package, ranging from C130's, to a P3. The box also houses full IFR and VFR charts, as well as approach plates.

In A Word: Skid marks?

Title: ## Airport 2000 (Vol.1)

Publisher/Developer: Wilco Publishing

Web Site: http://www.wilcopub.com

TecPilot Rating: ✦ ✦ ✦ ✦

Compatible with: FS98, FS2000

Requirements: Microsoft® Flight Simulator 98 or later - Pentium 133 MHz - 16 Mb Ram - 2x speed CD-ROM (for installation). Accelerator video cards highly recommended.

Description: The Airport 2000 series, has been one of the most successful scenery/plane/panel packages to come to flight simulator in a long time. While system demanding, the scenery's do feature very detailed buildings, and dynamic scenery. This package is not known for the aircraft that are supplied with it.

In A Word: A new era in FS design and technology.

Title: ## Airport 2000 (Vol.2)

Publisher/Developer: Wilco Publishing

Web Site: http://www.wilcopub.com

TecPilot Rating: ✦ ✦ ✦ ✦

Compatible with: FS98, FS2000

Requirements: PC CD-ROM. IBM PC & compatibles. Pentium 200 or higher. Windows 95/98. Microsoft Flight Simulator 2000. 32mb RAM. SVGA graphics card required. 2x CD-ROM or greater required

Description: The Airport 2000 Vol. 2, is the sequel to the highly popular Vol. 1. This time, Wilco brings you to new airports all over Europe, Asia, and the U.S.A., including, Chicago O'Hare, Boston Logan, Heathrow, and more.

In A Word: Getting good. What about the frame rates?

SCENERY

Title: ## Airport 2000 (Vol.3)

Publisher/Developer: Wilco Publishing

Web Site: http://www.wilcopub.com

TecPilot Rating: ✦ ✦ ✦ ✦

Compatible with: FS98, FS2000

Requirements: Pentium 600 MHz, 128 Mb RAM, 3D graphics accelerator card.

Description: Wilco's acclaimed series of scenery upgrades. Includes: 7 new airports, 3 new aircraft and 10 adventures. Airports: Berlin Tegel (Germany) / Paris Orly (France) / Kastrup Copenhagen (Denmark) / London Gatwick (United Kingdom) / Denver International / San Francisco International / Seattle Tacoma (USA) included.

In A Word: Frame rate killer. Loved by many, frowned upon by the few.

Title: ## Austria Pro

Publisher/Developer: Papa Tango

Web Site: http;//www.papatango.com

TecPilot Rating: ✦ ✦ ✦ ✦

Compatible with: FS98, CFS, FS2000

Requirements: Pentium 266 minimum (recommended), 64 Mb Ram, Microsoft Flight Simulator 98, CD Rom Drive, Soundblaster Compatible Sound Card, 2D/3D Card, Headphones or Speakers recommended.

Description: Papa Tango ventured into Austria, a country often left out in flight simulator add-ons, and took photo-realism to the extreme. The entire country is covered in elevated mesh terrain, as well as coloured photo-reaistic textures.

In A Word: Puts the D in Detailed!

Scenery

SCENERY

Title: ## Azores & Madeira

Publisher/Developer: Apollo Software

Web Site: http://www.apollosoftware.com

TecPilot Rating: ✦ ✦ ✦ ✦ ✦

Compatible with: FS95, FS98

Requirements: Pentium 266, 64 Mb Ram, CD Rom Drive. 2D/3D Card.

Description: Azores & Madeira scenery utilizes design techniques that recreated these beautiful islands in a way as never seen before on any flight simulation program. Designed by the same team that created the Europe I and Europe II scenery databases, this scenery package raised design standards to new heights.

In A Word: The pioneers of elevate mesh.

Title: ## Dangerous Airports

Publisher/Developer: Abacus

Web Site: http://www.abacuspub.com

TecPilot Rating: ✦ ✦ ✦ ✦ ✦

Compatible with: FS98

Requirements: Pentium 266 minimum (recommended), 64 Mb Ram, Microsoft Flight Simulator 98, CD Rom Drive. 2D/3D Card.

Description: Dangerous Airports is a collection of several airports located in some of the most challenging terrain in the world! The ten airports range in location from the Aleutian Islands to Iceland. The package also includes 4 interior and exterior aircraft models

In A Word: The IFR pilots delight!

Title: ## English Airports

Publisher/Developer: Just Flight / Barry Perfect

Web Site: http://www.justflight.com

TecPilot Rating: ✦ ✦ ✦ ✦ ✦

Compatible with: FS2000

Requirements: Pentium III 600 or higher, 128 Mb or more RAM, 32 Mb 3D Graphics accelerator card Monitor capable of 1024 x 768 resolution, 3D Sound card

Description: English Airports features Accurate Runways & Helipads - custom textured with tyre marks, Taxiways - that are accurately placed and sized with lines and night lighting, Buildings, - everything is there! Radar towers, hangars, office buildings, outdoor and multi-storey car parks, plus all piers are depicted complete with gates and air bridges and loads more!

In A Word: 6 of the best

Title: ### Europe 1

Publisher/Developer: Apollo Software

Web Site: http://www.apollosoftware.com

TecPilot Rating: ✦ ✦ ✦ ✦ ✦

Compatible with: FS98

Requirements: Pentium 266, 64 Mb Ram, CD Rom Drive. 2D/3D Card.

Description: Europe I is the original in a line of scenery databases to cover the European Continent. This first edition covers the counties of Germany, Austria, and Switzerland. New Palette techniques creates the green carpet of colors that are so well-known and cover the European countryside, in the scenery areas including hundreds of airports, cities, and villages... from Berlin to Bavaria.

In A Word: The start of a beautiful relationship

Title: ### Europe 2

Publisher/Developer: Apollo Software

Web Site: http://www.apollosoftware.com

TecPilot Rating: ✦ ✦ ✦ ✦ ✦

Compatible with: FS98

Requirements: Pentium 266, 64 Mb Ram, CD Rom Drive. 2D/3D Card.

Description: Europe II comes from the same design team that developed the best-selling Europe I scenery package. This addition to the continent includes the countries of France, Luxembourg, Belgium and Corsica. New scenery creation techniques and night lighting effects result in a stunning visual environment. Paris, the European countryside with its villages and cities come alive. The Europe II CD-ROM includes demos of other APOLLO products, as well as megabytes of video files.

In A Word: Will you marry me?

Title: ### Europe 3

Publisher/Developer: Apollo Software

Web Site: http://www.apollosoftware.com

TecPilot Rating: ✦ ✦ ✦ ✦ ✦

Compatible with: FS98, FS95, FS5

Requirements: Pentium 266, 64 Mb Ram, CD Rom Drive. 2D/3D Card.

Description: Europe III covers the area of Great Britain below the 54th parallel including Wales, England, Channel Islands and the Isles of Scilly. It includes more than 100 airports, buildings, all types of approach lights and airport signs, along with rivers, lakes, cities, mountains, forests. Even landmarks such as castles, factories and lighthouses are included.

In A Word: A classic in its time.

Title: **Flight Academy**

Publisher/Developer:	Papa Tango
Web Site:	http;//www.papatango.com
TecPilot Rating:	✦ ✦ ✦ ✦
Compatible with:	FS98
Requirements:	Pentium 266 minimum (recommended), 64 Mb Ram, Microsoft Flight Simulator 98, CD Rom Drive, Soundblaster Compatible Sound Card, 2D/3D Card, Headphones or Speakers recommended, 225Mb Disk Space.
Description:	Flight Academy is an original add-on that adds some life to several Eastern European airports. Features include working "follow me" cars, as well as other new scenery features, like snow, and runway lights that cast shadows!
In A Word:	Follow You Follow Me!

Title: **Geneva 2000**

Publisher/Developer:	Flight World
Web Site:	http://www.flightworld.org
TecPilot Rating:	✦ ✦ ✦ ✦ ✦
Compatible with:	FS2000, FS98
Requirements:	PC CD-ROM. IBM PC & compatibles. Pentium 200 or higher. Windows 95/98. Microsoft Flight Simulator 2000. 32mb RAM. SVGA graphics card required. 2x CD-ROM or greater required
Description:	Flight World dazzled up the area of and surrounding Geneva, in their Geneva scenery series. The scenery depicts the area in grave detail, including custom textured buildings, photo-realistic elements, and other various goodies that add to realism.
In A Word:	Will it be around long enough!

Title: **Germany VFR**

Publisher/Developer:	CR Software
Web Site:	http://www.cr-software.com/
TecPilot Rating:	✦ ✦ ✦ ✦ ✦
Compatible with:	FS98, FS2000
Requirements:	PC CD-ROM. IBM PC & compatibles. Pentium 266 or higher. Windows 95/98. Microsoft Flight Simulator 2000. 32mb RAM. SVGA graphics card required. 2x CD-ROM or greater required.
Description:	The emphasis of the CR Software VFR series was on realism. The CD includes mesh terrain, as well as photo-realistic textures of most all of Germany. Several airports are also re-done in greater detail than before.
In A Word:	Essential VFR

SCENERY

Title:

German Airports 1

Publisher/Developer: Aerosoft

Web Site: http://www.aerosoft.com

TecPilot Rating: ✦ ✦ ✦ ✦ ✦

Compatible with: FS98, FS95

Requirements: Pentium 200, 64 Mb Ram, Microsoft Flight Simulator 98, CD Rom Drive. 2D/3D Card.

Description: AeroSoft's first scenery in the GAP serise covers the airports of Munich, Augsburg, Stuttgart, Friedrichshafen, Nurenberg, Bayreuth, Dresden & Egelsbach in top quality. Each airport has custom textures and buildings.

In A Word: Where's the beer!

Title:

German Airports 2

Publisher/Developer: Aerosoft

Web Site: http://www.aerosoft.com

TecPilot Rating: ✦ ✦ ✦ ✦ ✦

Compatible with: FS98, FS2000

Requirements: Pentium 266, 64 Mb Ram, CD Rom Drive. 2D/3D Card.

Description: Includes the airports of Rhein/Main-Airport Frankfurt,Cologne/Bonn, Dortmund, Paderborn/Lippstadt, Munster/Osnabruck, Kassel, Hannover and Leipzig in an incredible top-quality!

In A Word: Unsurpassed detail

Title:

German Airports 3

Publisher/Developer: Aerosoft

Web Site: http://www.aerosoft.com

TecPilot Rating: ✦ ✦ ✦ ✦ ✦

Compatible with: FS98, FS2000

Requirements: Pentium 266, 64 Mb Ram, CD Rom Drive. 2D/3D Card.

Description: Includes the airports of Dusseldorf, Bremen, Hamburg, Kiel, Erfurt, Monchengladbach, Berlin-Tegel and Lubeck in an incredible top-quality!

In A Word: Setting standards

Title: ### German Airports 1 Edition 2001

Publisher/Developer: Aerosoft

Web Site: http://www.aerosoft.com

TecPilot Rating: ✦ ✦ ✦ ✦ ✦

Compatible with: FS2000

Requirements: Pentium 500, 64 Mb RAM, 3D Graphic Card

Description: German Airports 1 Edition 2001 is an update of the origin[...] the airports from the original product, in extremely high detail, as well as Innsbruck and Altenrhein as extras. GAP1 features "Action Scenery,"

In A Word: The dictionary definition of "a great update!"

Title: ### Great Britain & Ireland

Publisher/Developer: Just Flight / Barry Perfect

Web Site: http://www.aerosoft.com

TecPilot Rating: ✦ ✦ ✦

Compatible with: FS2000

Requirements: Pentium III 700 or higher, 128Mb or more RAM, 32Mb 3D graphics accelerator card, Monitor capable of 1024 x 768 resolution, 3D sound card

Description: Great Britain and Ireland contains over 280 highly detailed renditions of airports located around the British Isles. From international airports like Heathrow, Shannon and Manchester to smaller airports like Biggin Hill and Duxford, right down to local strips such as Kilkenny and Halfpenny Green. Everything in this collection has been carefully researched over 18 months in order to reproduce it at the highest level of detail.

In A Word: Well Done Barry!!

Title: ### Ireland

Publisher/Developer: Just Flight

Web Site: http;//www.justflight.com

TecPilot Rating: ✦ ✦ ✦ ✦ ✦

Compatible with: FS98

Requirements: PC CD-ROM. IBM PC & compatibles. Pentium 90 or higher. Windows 95/98. Microsoft Flight Simulator 98. 16mb RAM. SVGA graphics card required. 2x CD-ROM or greater required

Description: Ireland for FS98 adds new navigation beacons, cities, towns, and 30 new airports to the country. A special feature to Ireland, is virtual bird strikes! The scenery was designed by John Waller, one of the leading U.K. scenery designers.

In A Word: Bringing borders together

Title: ## Italy 98

Publisher/Developer: Lago

Web Site: http//www.lagoonline.com

TecPilot Rating: ✦ ✦ ✦ ✦

Compatible with: FS98. FS95

Requirements: Pentium 120 or higher microprocessor - 16Mb of total RAM - 50Mb of free space on the hard drive - Dual speed CD-ROM drive

Description: The standard sized box for Italy 98, houses 300,000 Km2 of detailed scenery. Including 140 monuments, accurate elevation data, even seasonal textures. All of 130 airports have been depicted in detail in the package too.

In A Word: Outdated but good in its day.

Title: ## Italy 2000

Publisher/Developer: Lago

Web Site: http//www.lagoonline.com

TecPilot Rating: ✦ ✦ ✦ ✦

Compatible with: FS2000

Requirements: Pentium II 350 - 32Mb of RAM - 60Mb of free space on the hard drive - CDROM Drive - Win95/98

Description: The included scenery uses the FS2000 default territory as its base, enhancing the exhisting airports by replacing the default buildings, adding new ones as well as the areas adiacent the airports. Also includes a set of tracks (moving aircraft) that fly the main procedures on all major and many minor airports.

In A Word: Bella!

Title: ## LA VFR

Publisher/Developer: CR Software

Web Site: http://www.cr-software.com/

TecPilot Rating: ✦ ✦ ✦

Compatible with: FS98, FS2000

Requirements: PC CD-ROM. IBM PC & compatibles. Pentium 266 or higher. Windows 95/98. Microsoft Flight Simulator 2000. 32mb RAM. SVGA graphics card required. 2x CD-ROM or greater required.

Description: The emphasis of the CR Software VFR series was on realism. The CD includes mesh terrain, as well as photo-realistic textures of most all of Los Angeles, and surrounding areas. Several airports are also re-done if greater detail than before.

In A Word: Compact and Bijou

SCENERY

Title:	**MegaScenery DownUnder - Victoria**
Publisher/Developer:	PC Aviator
Web Site:	http://www.pcaviator.com
TecPilot Rating:	✦ ✦ ✦ ✦ ✦
Compatible with:	FS98, CFS
Requirements:	Pentium 200 (recommended), 32 Mb Ram, Microsoft Flight Simulator 98, CD Rom Drive, SoundBlaster Compatible Sound Card, 2D/3D Card, Headphones or Speakers recommended.
Description:	Hundreds of miles of city, beach, and desert have been recreated by PC Aviator, in their Victoria Scenery. The package includes not only Melbourne, New Wales, and other cities, but also several Australian mountains and islands.
In A Word:	Better than shrimp on the barbie!

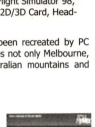

Title:	**Mexico Gold Pro (Vol I)**
Publisher/Developer:	FlyMex
Web Site:	http://www.flymex.com.mx/
TecPilot Rating:	✦ ✦ ✦ ✦ ✧
Compatible with:	FS2000
Requirements:	PC CD-ROM. IBM PC & compatibles. Pentium 333 or higher. Windows 95/98. Microsoft Flight Simulator 2000. 32mb RAM. SVGA graphics card required. 2x CD-ROM or greater required.
Description:	The airports contained in this volume extend from the central state of Zacatecas down to the airport of Chetumal which is located at the border line with Guatemala. This volume covers the air traffic control areas known as "Mexico center" and "Merida Center".
In A Word:	Ahh Tequila!

Title:	**Mexico Gold Pro (Vol II)**
Publisher/Developer:	FlyMex
Web Site:	http://www.flymex.com.mx/
TecPilot Rating:	✦ ✦ ✦ ✦ ✧
Compatible with:	FS2000
Requirements:	Microsoft Flight Simulator 2000, 64 MB RAM, Pentium 233 Mhz.
Description:	Covering North Mexico, Mexico Gold PRO Vol II includes, scenery, aircraft, approach charts, as well as new adventures and challenges inside the mexican airspace.
In A Word:	Give me another Tequila!

SCENERY

Title: Pacific Northwest

Publisher/Developer: PC Aviator

Web Site: http://www.pcaviator.com

TecPilot Rating: ✦ ✦ ✦ ✦ ✦

Compatible with: FS98, CFS

Requirements: Pentium 200 (recommended), 32 Mb Ram, Microsoft Flight Simulator 98, CD Rom Drive, SoundBlaster Compatible Sound Card, 2D/3D Card, Headphones or Speakers recommended.

Description: The Northwest United States gets a facelift in PC Aviator's Pacific Northwest. The scenery 65,000 miles with hi-resolution photo-realistic mesh terrain. Seattle, Portland and Vancouver are also included, and drawn in 5 meter per pixel resolution.

In A Word: Take your head out of the cockpit!

Title: Scenery Berlin 2000

Publisher/Developer: Aerosoft

Web Site: http://www.aerosoft.com

TecPilot Rating: ✦ ✦ ✦ ✦ ✦

Compatible with: FS98, FS95

Requirements: Pentium 200, 64 Mb Ram, Microsoft Flight Simulator 98, CD Rom Drive. 2D/3D Card.

Description: The metropolis Berlin with its 3 airports & Brandenburg with more than 20 airports. Includes many detailed objects.

In A Word: Detail? You better believe it!

Title: Scenery Bavaria

Publisher/Developer: Aerosoft

Web Site: http://www.aerosoft.com

TecPilot Rating: ✦ ✦ ✦ ✦ ✦

Compatible with: FS98, FS95

Requirements: Pentium 200, 64 Mb Ram, Microsoft Flight Simulator 98, CD Rom Drive. 2D/3D Card.

Description: This scenery was especially developed for the fan of visual navigation. It covers an area of more than 10.000 km² featuring all landmarks for most realistic VFR-navigation. All 52 airports and landing strips are included.

In A Word: Technically accurate!

SCENERY

Title: **Scenery Canary Islands**

Publisher/Developer: Aerosoft

Web Site: http://www.aerosoft.com

TecPilot Rating: ✦ ✦ ✦ ✦ ✦

Compatible with: FS98, FS95

Requirements: Pentium 200, 64 Mb Ram, Microsoft Flight Simulator 98, CD Rom Drive. 2D/3D Card.

Description: Take a flight over this photorealistic scenery of these very popular holiday islands and enjoy some varied mountain ranges and valleys in an unpresidented realism. All 7 islands with the 8 airports and many objects are waiting to be explored. Also 2 aircraft are included: Boeing 767-200 (Spanair) and Airbus A320 (LTU).

In A Word: Tweet !

Title: **Scenery China**

Publisher/Developer: Aerosoft

Web Site: http://www.aerosoft.com

TecPilot Rating: ✦ ✦ ✦ ✦

Compatible with: FS98, FS95

Requirements: Pentium 200, 64 Mb Ram, Microsoft Flight Simulator 98, CD Rom Drive. 2D/3D Card.

Description: Huge scenery covering approx. 9.5 million km². 33 airports and many highlight like the Great Wall, Beijing. 2 adventures and 10 highly detailed typical aircraft included.

In A Word: Big country, big add-on

Title: **Scenery Rhein/Ruhr**

Publisher/Developer: Aerosoft

Web Site: http://www.aerosoft.com

TecPilot Rating: ✦ ✦ ✦ ✦ ✦

Compatible with: Fly! , Fly!2K

Requirements: Pentium 200, 64 Mb Ram, CD Rom Drive. 2D/3D Card.

Description: This incredible scenery-add-on lets you experience the area of the so called Rhur-Gebiet. The densest-populated area in Germany. The great number of airports and landmarks make this add-on a must for every VFR-Pilot

In A Word: Where's my camera!

Title:

Scenery Singapore

Publisher/Developer: Lago

Web Site: http://www.lagoonline.com

TecPilot Rating: ✦✦✦✦✦

Compatible with: FS98

Requirements: Pentium 266 minimum (recommended), 64 Mb Ram, Microsoft Flight Simulator 98, CD Rom Drive. 2D/3D Card.

Description: Includes several airports and airbases: Changi International, Paya Lebar, Seletar Airport and Tengah Airbase. Over 400 buildings and objects are available scattered around the island mostly in the East, the City, Woodlands and Johore Bahru.

In A Word: Good reviews in its day

SCENERY

Title:

Scenery Tokyo

Publisher/Developer: Lago

Web Site: http://www.lagoonline.com

TecPilot Rating: ✦✦✦✦✦

Compatible with: FS98, CFS

Requirements: Pentium 266 minimum, 64 Mb Ram, , CD Rom Drive. 2D/3D Card.

Description: The Virtual Tokyo, simulates the city's major, and minor landmarks, as well as the ground bellow. Accurate elevation data, is combined with high resolution satellite images to give the scenery a realistic feeling, when looking out the window.

In A Word: Small, yet tantalizing.

Title:

Sky Ranch

Publisher/Developer: Abacus

Web Site: http://www.abacuspub.com

TecPilot Rating: ✦✦✦✦✦

Compatible with: FS2002

Requirements: Pentium 600, 64 MByte RAM, 3D Graphic Card

Description: See the world from a different perspective. Grassy Meadows. Sky Ranch (UT47) is located in southern Utah. It maximizes the full potential of FS2002's scenery engine . A little niche but someone will love using it as a base to explore the wonderful region.

In A Word: Bonanza, or was it Little House on the Prairie?

Title: ## Swizerland 2

Switzerland 2

Publisher/Developer: Flight and Cockpit

Web Site: http://www.pcflight.ch/

TecPilot Rating: ✦ ✦ ✦ ✦

Compatible with: FS2000

Requirements: PC CD-ROM. IBM PC & compatibles. Pentium 200 or higher. Windows 95/98. Microsoft Flight Simulator 2000. 32mb RAM. SVGA graphics card required. 2x CD-ROM or greater required

Description: Switzerland 2 is scenery covering all of Switzerland, including the Alps. The terrain is accurate mesh, covered with aerial images. The package includes accurately designed airports throughout the country.

In A Word: The hills are alive!

Title: ## Thailand, Malaysia, Singapore

Publisher/Developer: Aerosoft

Web Site: http://www.aerosoft.com

TecPilot Rating: ✦ ✦ ✦ ✦

Compatible with: FS98, FS95, FS5

Requirements: Pentium 200, 64 Mb Ram, Microsoft Flight Simulator 98, CD Rom Drive. 2D/3D Card.

Description: This scenery covers more than 840.000 km² of Thailand, Singapore and western Malaysia. Considerable attention was paid to the visual representation of this typical asian landscape with all its exotic colours and beauty. Contains over 50 airports, many objects and other scenery details.

In A Word: Where's the plane?

Title: ## The Triangle

Publisher/Developer: Lago

Web Site: http//www.lagoonline.com

TecPilot Rating: ✦ ✦ ✦ ✦

Compatible with: FS98

Requirements: Pentium 266 minimum (recommended), 64 Mb Ram, Microsoft Flight Simulator 98, CD Rom Drive. 2D/3D Card.

Description: LAGO sport's its "iZone interactive scenery" in their The Triangle package. The scenery, which covers the mysterious Bermuda Triangle, allows you to play various "games", as well as use flight simulator, as a boat sim.

In A Word: Weird!

a-Z
Boxed Add-on Products
Utilities

Title: **Aircraft Animator**

Publisher/Developer: Abacus

Web Site: http://www.abacuspub.com

TecPilot Rating: ✦ ✦ ✦ ✦ ✦

Compatible with: FS2000, FS98, CFS

Requirements: Pentium 266 minimum (recommended), 64 Mb Ram, CD Rom Drive. 2D/3D Card.

Description: AA was quite a treat to aircraft developer's ears when it arrived. The program, allows even the less than average aircraft developer, to add dynamic surfaces to the plane (gear, flaps, spoilers, propeller, and shining landing lights)

In A Word: Sophisticate, yet simple.

Title: **Aircraft Factory 99**

Publisher/Developer: Abacus

Web Site: http://www.abacuspub.com

TecPilot Rating: ✦ ✦ ✦ ✦ ✦

Compatible with: FS98, CFS, FS2000

Requirements: Pentium II 300 PC, Windows 98, 98, 64 Mb Ram, CD ROM Drive, Sound Card.

Description: Aircraft Factory is the simplest way for a person to create an FS aircraft. It is an easy solution to plotting every point by hand. The program also includes at detailed tutorial to help you get started.

In A Word: Up in the air faster then ever!

Title: **Airport & Scenery Designer V2.1**

Publisher/Developer: Abacus

Web Site: http://www.abacuspub.com

TecPilot Rating: ✦ ✦ ✦ ✦

Compatible with: FS98, FS2000, CFS

Requirements: Pentium 266 minimum (recommended), 64 Mb Ram, Microsoft Flight Simulator 98, CD Rom Drive. 2D/3D Card.

Description: A&SD is by far one of the most, if not the most popular scenery designers for flight simulator. It's easy to use graphical interface is one possible attribute to it's success. A&SD is loaded with tools, from pilot controlled lighting to elevated mesh terrain scenery.

In A Word: Look, there's our house!

Title: **Creatools**

Publisher/Developer: Flightworld

Web Site: http://www.flightworld.org

TecPilot Rating: ✦ ✦ ✦ ✦ ✦

Compatible with: FS98, FS2000

Requirements: PC CD-ROM. IBM PC & compatibles. Pentium 200 or higher. Windows 95/98. 32mb RAM. SVGA graphics card required. 2x CD-ROM or greater required.

Description: Package of scenery creation utilities. The package includes tools from EZBuilding, which allows you to create custom buildings "in minutes," to EZDots which creates various night lighting for flight simulator.

In A Word: Complicated and crammed

Title: **Custom Panel Designer**

Publisher/Developer: Abacus

Web Site: http://www.abacuspub.com

TecPilot Rating: ✦ ✦ ✦ ✦ ✦

Compatible with: FS98, FS2000, CFS, CFS2

Requirements: Pentium 266 minimum (recommended), 32 Mb Ram,CD Rom Drive. 2D/3D Card.

Description: Custom Panel Designer, is a simple to use Flight Sim panel designer. After loading a bitmap panel image, you simply drag, and size gauges into there appropriate positions.

In A Word: Another easy alternative

Title: **Final Approach**

Publisher/Developer: Just Flight

Web Site: Http;//www.justflight.com

TecPilot Rating: ✦ ✦ ✦ ✦ ✦

Compatible with: FS95, FS98, FS2000, FU3, Fly!, ProPilot

Requirements: 486/50 MHz PC, 8 MB of RAM, Windows 95 or 98, Printer required for print function, CD ROM drive.

Description: Final Approach 2000 is a powerful, Windows-based approach chart atlas & designer for use with Microsoft Flight Simulator, Fly!, Flight Unlimited III, Pro Pilot, 747 PS1 Precision Simulator or any other realistic general-aviation flight simulator. This 2000 version builds on a heritage of success. Create your own approach charts – Includes a multitude of ready-to-go navigation plate symbols and functions to make the creation of realistic Approach, STARS and SIDS plates easy.

In A Word: The premiere chart program.

Utilities

Title: Flight Director 99

Publisher/Developer: Just Flight

Web Site: Http;//www.justflight.com

TecPilot Rating: ✦ ✦ ✦ ✦ ✧

Compatible with: FS98, FS2000

Requirements: 64 Mb ram. (Internet connection is required for downloading live weather data). A 3D video card capable of runing a 3D window within window.

Description: Integrates Real Weather, Flight Planning, GPS, EFIS, Weather Radar, and ATC in the ultimate companion to Microsoft Flight Simulator 98.

In A Word: Nuts and Bolts!

Title: Flight Planner 98

Publisher/Developer: Aerosoft

Web Site: http://www.aerosoft.com

TecPilot Rating: ✦ ✦ ✦ ✧ ✦

Compatible with: FS98, FS2000

Requirements: Pentium 200, 64 Mb Ram, CD Rom Drive. 2D/3D Card.

Description: Create flight plans both automatically and manually. Choose great circle calculation for long-distance flights. Include intersections in your flight plan. Show a Jeppesen-style approach chart. Calculate fuel consumption, alternate airports, etc. Look at a map displaying.

In A Word: Simple, but useful.

Title: Flight Shop

Publisher/Developer: Apollo

Web Site: http://www.apollosoftware.com

TecPilot Rating: ✦ ✦ ✦ ✧ ✦

Compatible with: FS98, FS95

Requirements: Pentium 200, 32 Mb Ram, CD Rom Drive. 2D/3D Card.

Description: Flight Shop is the original airplane creator for FS. It allows you to draw a wireframe of a plane, the paint and create it's flight dynamics. Then your plane is ready to fly.

In A Word: Only used for nostalgia.

Title: FS Clouds 2000

Publisher/Developer: Flight 1

Web Site: http://www.flight1.com

TecPilot Rating: ✦ ✦ ✦ ✦ ✦

Compatible with: FS2000

Requirements: PC CD-ROM. IBM PC & compatibles. Pentium 200 or higher. Windows 95/98. 64mb RAM. SVGA graphics card required. 2x CD-ROM or greater required

Description: Fly through practically unlimited skyscapes using the more than 60 new cloud components. See new dimensions in the sky as you fly past clouds that appear much more vaporous and volumetric.

In A Word: Wafting through flight sim instead of hitting blocks!

Title: FS Design Studio

Publisher/Developer: Abacus

Web Site: http://www.abacuspub.com

TecPilot Rating: ✦ ✦ ✦ ✦

Compatible with: FS2000, CFS

Requirements: PC CD-ROM. IBM PC & compatibles. Pentium 200 or higher. Windows 95/98. 32mb RAM. SVGA graphics card required. 2x CD-ROM or greater required

Description: FSDesign studio is a very powerful tool that allows you to create your own custom scenery macros. These macros can be placed into scenery by either inputing Lat. and Lon., or you can place it in the FS world using Airport & Scenery Designer.

In A Word: Get the "pro" version

Title: FS Design Studio Pro

Publisher/Developer: Abacus

Web Site: http://www.abacuspub.com

TecPilot Rating: ✦ ✦ ✦ ✦ ✦

Compatible with: FS2000, CFS

Requirements: PC CD-ROM. IBM PC & compatibles. Pentium 200 or higher. Windows 95/98. 32mb RAM. SVGA graphics card required. 2x CD-ROM or greater required

Description: Create Superb 3D Scenery Objects and Aircraft. FS Design Studio gives you the easiest way yet to create 3D scenery objects and flyable aircraft. With four simultaneous views, you can immediately see your designs in real time as they'll appear in FS2000.

In A Word: A true a breakthrough in 3D design.

Title:	**FS Flightbag**
Publisher/Developer:	Abacus
Web Site:	http://www.abacuspub.com
TecPilot Rating:	✦ ✦ ✧ ✧ ✧
Compatible with:	FS98, FS2000
Requirements:	Pentium 266 minimum (recommended), 32 Mb Ram, CD Rom Drive. 2D/3D Card.
Description:	FS Flightbag is the first step in becoming an IFR pro. The GPS and other utilities included are easy to use and simple to understand.
In A Word:	No use these days.

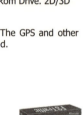

Title:	**FS Traffic**
Publisher/Developer:	Lago
Web Site:	http://www.lagoonline.com
TecPilot Rating:	✦ ✦ ✦ ✦ ✦
Compatible with:	FS2000, FS98
Requirements:	PC CD-ROM. IBM PC & compatibles. Pentium 200 or higher. Windows 95/98. 32mb RAM. SVGA graphics card required. 2x CD-ROM or greater required.
Description:	Comparing it with the standard dynamic traffic you are used until now, you will find out that the traffic generated by FSTraffic is much more powerful, customisable and will add a new dimension to your flights.
In A Word:	Unbelievably busy!

Title:	**FS Action Scenery** (Animated Objects)
Publisher/Developer:	Abacus
Web Site:	http://www.abacuspub.com
TecPilot Rating:	✦ ✦ ✦ ✧ ✧
Compatible with:	FS2000
Requirements:	Pentium 300 minimum, 64 Mb Ram, CD Rom Drive. 2D/3D Card
Description:	Using FS Action! Scenery you add this missing element to all your flights. In just a few minutes you can animate any of your "static scenery areas" with this easy-to-use tool. Designer Louis Sinclair makes it easy for you to add dynamic scenery everywhere.
In A Word:	Movin' on up!

UTILITIES

Title: Instant Airplane Maker

Publisher/Developer: Abacus

Web Site: http://www.abacuspub.com

TecPilot Rating: ✦ ✦ ✦ ✦ ✦

Compatible with: FS2000, FS98

Requirements: Pentium 266 minimum (recommended), 32 Mb Ram, CD Rom Drive. 2D/3D Card.

Description: Instant Plane Maker allows you to paint and customize your own FS aircraft. This powerful program is so simple anyone can do a custom paint job in just hours!

In A Word: Airlines in a matter of minutes!

Title: Precision Pilot

Publisher/Developer: Just Flight

Web Site: http://www.justflight.com

TecPilot Rating: ✦ ✦ ✦ ✦ ✦

Compatible with: All Simulations

Requirements: P266 MHz or higher CPU. 32MB RAM. 3D graphics accelerator card capable of 1024x768 resolution and 24 bit colour.

Description: A stand-alone navigation and approach trainer package that helps all simulation aviators progress their skills and fly like the pros. Provides an interactive and graded tutorial system for pilots to learn the methodology of instrument-only navigation and associated skills. Learn what an NDB is, how to execute a holding pattern or perform a professional ILS approach and much more.

In A Word: North is THAT way!

Title: Private Pilot Training

Publisher/Developer: Abacus

Web Site: http://www.abacuspub.com

TecPilot Rating: ✦ ✦ ✦ ✦ ✦

Compatible with: FS2000

Requirements: PC CD-ROM. IBM PC & compatibles. Pentium 200 or higher. Windows 95/98. 32mb RAM. SVGA graphics card required. 2x CD-ROM or greater required.

Description: Private Pilot Training is one of the first tools that allows you to use FS2000 as a serious flight trainer. The program gives you individualized instruction as would a real CFI. The interface for the lessons was is very attractive, and easy to use. One program will teach you all the basics of VFR.

In A Word: Great lessons, without the huge price.

Title: **Ultimate Airlines**

Publisher/Developer: Flight One

Web Site: http://www.flight1.com

TecPilot Rating: ✦ ✦ ✦ ✦

Compatible with: FS2000

Requirements: Pentium 600, 64 MByte RAM, CDROM, 3D Graphic Card

Description: Ultimate Airlines will provide time the time tables for most all airline flights in the whole world! The program allows realistic flight plans to be created, and be exported to AETI's ProFlight, to be turned into voice ATC flights.

In A Word: The ultimate airline operations utility!

a-z
Download Add-on Products
Adventures

Title: **Adventure Designer Pack**

Publisher/Developer:	AriFlight
Web Site:	http://www.ari.de/
TecPilot Rating:	✦ ✦ ✦ ✦ ✦
Compatible with:	FS98
Requirements:	486 processor, Windows(r)95 or NT, 16Mb RAM
Description:	This program allows you to make your own FS98 adventures. You can also choose to have your Co Pilot fly the trip for you.
In A Word:	It's a great program, if you're still in 1997.

Title: **Adventure IDE**

Publisher/Developer:	John Hnidec
Web Site:	http://www.simsystems.com.au/
TecPilot Rating:	✦ ✦ ✦ ✦ ✦
Compatible with:	FS98, FS2000
Requirements:	IBM PC & compatibles. Pentium 200 or higher. Windows 95/98. 32mb RAM. SVGA graphics card required.
Description:	AdventureIDE is a powerful, professional environment for creating and managing Microsoft Flight Simulator adventures that you create. It is a self contained application which provides all the tools and facilities required to simplify the task of adventure creation. For programmers out there, it is similar to the Microsoft Visual Studio or the Inprise IDE development environments.
In A Word:	Bring your calculator!

Title: **FS2000 ATC**

Publisher/Developer:	Stephen Coleman
Web Site:	http://www.stephencoleman.com/fs2000.html
TecPilot Rating:	✦ ✦ ✦ ✦ ✦
Compatible with:	FS98, FS2000
Requirements:	Pentium 300 minimum (recommended), 64 Mb Ram, Soundblaster Compatible Sound Card, 2D/3D Card, Headphones or Speakers recommended.
Description:	Powerful adventure creator that also adds audio ATC to several portions of the flight including checklists.
In A Word:	Seen better— however, a cheaper alternative to some commercial products.

Title: Perfect Flight 2000

Publisher/Developer: Marco Martini

Web Site: http://digilander.iol.it/fs2000/pf2kgen.htm

TecPilot Rating: ✦ ✦ ✦ ✦ ✦

Compatible with: FS2000

Requirements: Pentium 600, 128 MByte RAM, 3D Graphic Card

Description: Perfect Flight 2000 is a novel adventure generator. This program allows you to set V1, V2, and VR speeds, as well as vary the weather conditions along the flight path. The program also handles SID, and STAR procedures. A GPSW is also included.

In A Word: A perfect flight!

Title: Radar Contact v2.1

Publisher/Developer: JDT LLC

Web Site: http://www.flightsimmers.net/radarcontact/

TecPilot Rating: ✦ ✦ ✦ ✦ ✦

Compatible with: FS98, FS2000

Requirements: Pentium 500, 64 MByte RAM, 3D Graphic Card (Viper 770 or similar)

Description: Adventure generator for Flight Simulator. This software takes flight plans generated by third party flight planners and generates spoken ATC which will provide the most realistic flight experience, adhering to the FAA rules. JDT LLC is owned and operated by John Dekker and Doug Thompson.

In A Word: A legend in realistic ATC adventures.

ADVENTURES

a-Z
Download Add-on Products
Aircraft

Title: ## Abacus DC-3

Publisher/Developer: Abacus

Web Site: http://www.flightsimdownloads.com

TecPilot Rating: ✦ ✦ ✦ ✦ ✧

Compatible with: FS2000, FS98,

Requirements: Pentium II 300 PC, Windows 95/ 98, 64 Mb Ram, Sound Card

Description: The DC-3 is considered to be the "classic" airliner of all time. For years it was the workhorse of many commercial airline companies after having served in World War II as the C-47. Luckily for us, there are still many DC-3s flying today.

In A Word: Feels real and it's FREE!

Title: ## A-10 Thunderbolt II

Publisher/Developer: Abacus

Web Site: http://www.flightsimdownloads.com

TecPilot Rating: ✦ ✦ ✦ ✦ ✧

Compatible with: CFS2, FS2000, FS2002

Requirements: 600MHz Processor, 128Mb RAM, Sound Card, 3D Accelerator Card

Description: This A-10 Thunderbolt comes with a very detailed exterior model, as well as a 3D, look around panel. The A-10, or "warthog" as it is affectionately known, was designed for FS002 by Dave Eckert. This A-10 can be downloaded, and used for 7 days before registering is required.

In A Word: Fast and furious!

Title: ## Beech Baron

Publisher/Developer: Abacus

Web Site: http://www.flightsimdownloads.com

TecPilot Rating: ✦ ✦ ✦ ✧ ✧

Compatible with: FS2000 Only

Requirements: Pentium II 300 PC, Windows 95/ 98, 64 Mb Ram, Sound Card

Description: The Baron is one of aviation's favorite twin and Terry Hill has captured this plane with this slick rendition. It combines a fabulous visual model with full moving parts, accurate flight characteristics and a crisp, clean instrument panel.

In A Word: Nice but a bit stiff.

Title:　　　　　　　### Boeing 747

Publisher/Developer:　Abacus

Web Site:　　http://www.flightsimdownloads.com

TecPilot Rating:　✦ ✦ ✦ ✧ ✧

Compatible with:　FS2000, FS98

Requirements:　Pentium II 300 PC, Windows 95/ 98, 64 Mb Ram, Sound Card

Description:　This "try before you buy" 747 includes a very detailed exterior, as well as a high resolution instrument panel. Also included are sounds.

In A Word:　Great treat for a low price.

Title:　　　　　　　### Boeing 777-200

Publisher/Developer:　Abacus

Web Site:　　http://www.flightsimdownloads.com

TecPilot Rating:　✦ ✦ ✦ ✧ ✧

Compatible with:　FS2000, FS98

Requirements:　Pentium II 300 PC, Windows 95/ 98, 64 Mb Ram, Sound Card

Description:　The Abacus 777 features a very round body, painted in the United Airlines color. The plane also includes a panel that has alll the 777's advanced avionics. It also includes moving parts.

In A Word:　Doesn't match other 777's.

Title:　　　　　　　### Cessna 310

Publisher/Developer:　Abacus

Web Site:　　http://www.flightsimdownloads.com

TecPilot Rating:　✦ ✦ ✦ ✧ ✧

Compatible with:　CFS2, FS2000, FS2002

Requirements:　600MHz Processor, 128Mb RAM, Sound Card, 3D Accelerator Card

Description:　This orange painted Cessna 310 from Abacus includes an exterior model, panel, as well as sounds. You can download and try the plane for 7 days before buying it.

In A Word:　Not that hot

Title: ## Concorde SST

Publisher/Developer: Abacus

Web Site: http://www.flightsimdownloads.com

TecPilot Rating: ✦ ✦ ✦ ✦ ✧

Compatible with: FS2000, FS98, FS2002

Requirements: Pentium II 300 PC, Windows 95/ 98, 64 Mb Ram, Sound Card

Description: This Concorde is like no other for FS98. It has a very detailed exterior, as well as a very complete panel. The plane is available in a British Airway livery.

In A Word: Fun, but only if your using FS98.

Title: ## F4U-5 Corsair

Publisher/Developer: Abacus

Web Site: http://www.flightsimdownloads.com

TecPilot Rating: ✦ ✦ ✦ ✦

Compatible with: CFS, FS2000

Requirements: Pentium II 300 PC, Windows 95/ 98, 64 Mb Ram, Sound Card

Description: Alain L'Homme has created this striking replica of the Corsair. This unique aircraft features moving parts: rotating propeller, landing gear that retract as smooth-as-silk; flaps that drop down, sliding transparent canopy and engine cowling, moving ailerons and elevator. This Corsair is tuned for Combat Flight Simulator only and includes a damage profile.

In A Word: Pacifically Good!!

Title: ## GeeBee Racer

Publisher/Developer: Abacus

Web Site: http://www.flightsimdownloads.com

TecPilot Rating: ✦ ✦ ✦ ✦

Compatible with: FS2000 Only

Requirements: Pentium II 300 PC, Windows 95/ 98, 64 Mb Ram, Sound Card

Description: Graphic artist Scott Nix recreates this spectacular reproduction the famous GeeBee, a craft from a bygone era. The paint scheme is unlike any that you're ever seen before. This truly amazing plane is made using FS Design Studio. Includes Panel.

In A Word: Fabulous and Fun!

AIRCRAFT

Title: **Giants of the Sky**
Gold Edition: The 747-400

Publisher/Developer: Flight1

Web Site: http://www.flight1.com

TecPilot Rating: ✦ ✦ ✦ ✦ ✦

Compatible with: FS2000

Requirements: The new panel requires a fair amount of computing power, so at least a Pentium II 400 is recommended, with 128MB of RAM. You should also have at least 100 MB of spare disk space, not including the install files you download.

Description: Giants of the Sky is yet another excellent 747. It includes a panoramic panel, as well as a highly detailed exterior, as well as stereo sounds. 12 different airlines are depicted in this package.

In A Word: One of the best 747's flying!

Title: **Glassair III**

Publisher/Developer: Abacus

Web Site: http://www.flightsimdownloads.com

TecPilot Rating: ✦ ✦ ✦ ✦ ✦

Compatible with: FS2000

Requirements: Pentium II 300 PC, Windows 95/ 98, 64 Mb Ram, Sound Card

Description: The Glassair is a high performance kitplane. But for those who'd rather not spend the time to build the kit, Terry Hill has assembled it for you using the FS Design Studio tool. This cool plane can easily reach 200 knots and fly 1200+ miles.

In A Word: Done, and done again!

AIRCRAFT

Title: **Greatest Airliners: 737-400**

Publisher/Developer: Flight1

Web Site: http://www.flight1.com

TecPilot Rating: ✦ ✦ ✦ ✦ ✦

Compatible with: FS2000

Requirements: 600MHz Processor, 128Mb RAM, Sound Card, 3D Accelerator Card

Description: The Greatest Airliners: 737-400 was perhaps the most anticipated flight sim release in memory. The package includes the most detailed panel ever. It included the main forward panel, as well as overhead, and FMC views. The entire panel was made from actual cockpit photos. The airplane its self comes in several different liveries, and includes a program that allows you to create your own. The 737 features many complex components of the real 737, such as the Flap Load Relief system, and the flight announcement system.

In A Word: Not just one of the greatest airliners, one of the greatest add-ons!

Title: ## Messerschmitt Me262

Publisher/Developer: Abacus

Web Site: http://www.flightsimdownloads.com

TecPilot Rating: ✦ ✦ ✦ ✦ ✦

Compatible with: FS2000, CFS

Requirements: Pentium II 300 PC, Windows 95/ 98, 64 Mb Ram, Sound Card

Description: The "secret weapon" that never quite made it into production towards the end of WWII. One of the first "jets" which served as an example for later aircraft. With moving parts, custom sounds and panel and ready for combat missions.

In A Word: Fun, and challenging.

Title: ## Mooney M20R Ovation

Publisher/Developer: Abacus

Web Site: http://www.flightsimdownloads.com

TecPilot Rating: ✦ ✦ ✦ ✦ ✦

Compatible with: FS2000

Requirements: Pentium II 300 PC, Windows 95/ 98, 64 Mb Ram, Sound Card

Description: This slick plane is a "personal airliner" from Mooney, a name that is synonymous with fast, high performers. The newest model Ovation cruises at up to 190 knots and will take you an incredible 1300 miles.

In A Word: Slick livery—average handling—bad panel.

Title: ## Piper Aztec

Publisher/Developer: Abacus

Web Site: http://www.flightsimdownloads.com

TecPilot Rating: ✦ ✦ ✦ ✦ ✦

Compatible with: CFS2, FS2000, FS2002

Requirements: 600MHz Processor, 128Mb RAM, Sound Card, 3D Accelerator Card

Description: The Piper PA-23 is more commonly known as the Aztec. This six-place twin was originally launched in the early 1960's. With two-250 hp engines, it can cruise at 205 mph at 21,000 feet and travel more than 1150 miles. Add-on maker Dave Eckert has designed this Aztec for FS2000 users with the help of FS Design Studio.

In A Word: Naughty, but nice!

Title: ## Piper Cub

Publisher/Developer: Abacus

Web Site: http://www.flightsimdownloads.com

TecPilot Rating: ✦ ✦ ✦ ✧ ✧

Compatible with: FS2000, FS2002, FS98 and CFS

Requirements: Pentium II 300 PC, Windows 95/ 98, 64 Mb Ram, Sound Card

Description: Since its introduction in the mid-1940's, the Piper Cub has become synonomous with general aviation. Designer Matt Garry is a Piper Cub owner himself so he has experienced the joy of flying this "vintage" aircraft with the original "old style" cockpit.

In A Word: Puke Colour—pavement pizza!

Title: ## Piper Comanche 250

Publisher/Developer: Abacus

Web Site: http://www.flightsimdownloads.com

TecPilot Rating: ✦ ✦ ✦ ✧ ✧

Compatible with: FS2000 Only

Requirements: Pentium II 300 PC, Windows 95/ 98, 64 Mb Ram, Sound Card

Description: The popular Comanche is a retractable single that can whiz along at 155 knots with a range of 600 nautical miles. Experience flying Terry Hill's rendition of this great aircraft.

In A Word: Very nice— slick—Good Panel!

AIRCRAFT

Title: ## Piper Navajo

Publisher/Developer: Abacus

Web Site: http://www.flightsimdownloads.com

TecPilot Rating: ✦ ✦ ✦ ✧ ✧

Compatible with: FS2000, FS2002

Requirements: Pentium II 300 PC, Windows 95/ 98, 64 Mb Ram, Sound Card

Description: Terry Hill and Jim Rhoads have teamed up to produce this beauty using our very own new cutting edge tool FS Design Studio. This six/eight place twin is a popular business plane. It can carry enough fuel for a 800 to 1000 mile trip at 215 mph. This is a "next generation" aircraft .

In A Word: Nice aircraft model—generic looking panel.

Title: ## P-51 Dakota Kid Mustang

Publisher/Developer: Abacus

Web Site: http://www.flightsimdownloads.com

TecPilot Rating: ✦ ✦ ✦ ✧

Compatible with: FS2000, FS98, CFS

Requirements: Pentium II 300 PC, Windows 95/ 98, 64 Mb Ram, Sound Card

Description: This P-51, nicknamed "The Dakota Kid" features moving parts: rotating propeller, smoothly retracting landing gear; large drop-down flaps, a sliding canopy and more. This P-51 is ready to do battle with Combat Flight Simulator and includes a damage profile. In short it's ready for combat missions! (and also flies with FS2000)

In A Word: Yeeeee ha!!

Title: ## P-61 Black Widow

Publisher/Developer: Abacus

Web Site: http://www.flightsimdownloads.com

TecPilot Rating: ✦ ✦ ✦ ✧

Compatible with: CFS2, FS2002, FS2000

Requirements: 600MHz Processor, 128Mb RAM, Sound Card, 3D Accelerator Card

Description: This P-61B Black Widow design is based on a restoration project currently underway at the Mid-Atlantic Air Museum in Reading, PA. Designed as the only night fighter of World War II by Northrop, the P-61 had a distinighed career against both the German Luftwaffe and Japanese IAF

In A Word: Wow and wow again!

Title: ## Royal Navy Aviation Collection

Publisher/Developer: VFR Scenery

Web Site: http://www.vfrscenery.com/

TecPilot Rating: ✦ ✦ ✦ ✧

Compatible with: FS98, FS2000, CFS and CFS2

Requirements: 600MHz Processor, 128Mb RAM, Sound Card, 3D Accelerator Card

Description: The Royal Navy Aviation Collection includes 30 of Great Britain's finest planes. Aircraft such as the De Havilland Sea Venom, and the F4F Wildcat have been included. The collection also comes with several CFS missions, just as "Attack on the Bismarck" and "Operation Fuller"

In A Word: England's finest, by one of flight sim's finest!

AIRCRAFT

Title: **Supermarine Spitfire**

Publisher/Developer: Abacus

Web Site: http://www.flightsimdownloads.com

TecPilot Rating: ✦ ✦ ✧ ✧ ✧

Compatible with: FS2000, CFS

Requirements: Pentium II 300 PC, Windows 95/ 98, 64 Mb Ram, Sound Card

Description: The "SPIT" was a top fighter during WWII and many consider it to be the most important aircraft of the 40s. This Mk IX, designated as the "MX-Y" has moving parts including the rotating propeller, smoothly retracting landing gear, two-part flaps and sliding canopy.

In A Word: We've seen better

Title: **The Concorde Experience**

Publisher/Developer: Paul Delaney

Web Site: http://www.users.globalnet.co.uk/~pdelaney

TecPilot Rating: ✦ ✦ ✦ ✧

Compatible with: FS98, FS2000

Requirements: Pentium II 300 PC, Windows 95/ 98, 64 Mb Ram, Sound Card

Description: This package will add a British Airways liveried Concorde (G-BOAC) to FS98 or FS2000. Includes a fairly accurate flight model (within the limitations of Flight Simulator), a realistic instrument panel and a fully functioning Inertial Navigation System. For FS98 users, scenery for Bahrain International Airport is also included.

In A Word: Past it!

Title: **TLK-39C Pilot Training Device**

Publisher/Developer: Captain Simulations

Web Site: http://www.web-captain.com

TecPilot Rating: ✦ ✦ ✦ ✦

Compatible with: FS2000, CFS2

Requirements: 600MHz Processor, 128Mb RAM, Sound Card, 3D Accelerator Card

Description: Captain Simulation's TLK-39C Training Device, based on a Ukrainian L-39 Albatros. It include an extremely detailed aircraft model, as well as a 3D panel. The panel is equipped with more then six dozen gauges and switches. Detailed scenery of the Konotop Air Force Base, featuring realistic textures, and dynamic scenery, is also included. The English edition of the package will come with approach, as well as IFR and VFR charts, all of which are USSR Air Force issue.

In A Word: A different approach!

Title: ## Twin Bonanza

Publisher/Developer: Abacus

Web Site: http://www.flightsimdownloads.com

TecPilot Rating: ✦ ✦ ✦ ✧ ✧

Compatible with: FS2000, FS2002

Requirements: 600MHz Processor, 128Mb RAM, Sound Card, 3D Accelerator Card

Description: This cartoon-ish looking 1960s Beach includes transparent windows, and moving control surfaces, as well as an impressive panel.

In A Word: Proof that something good did come out of the 60's.

LAST MINUTE ADDITIONS

Title: ## Piper Seneca

Publisher/Developer: Phoenix Simulation

Web Site: http://www.phoenix-simulation.co.uk

TecPilot Rating: ✦ ✦ ✦ ✦ ✦

Compatible with: FS2002

Requirements: 800MHz Processor, 128Mb RAM, Sound Card, 3D Accelerator Card

Description: One of the first aircraft additions for FS2002. Includes opening doors, reflection effects, rotating wheels among other stunning details.

In A Word: Simulation has started to get frighteningly realistic. Phoenix, don't get complacent. you are getting too good! Keep it up!

a-Z
Download Add-on Products
Scenery

Title: ## Amsterdam

Publisher/Developer: AriFlight

Web Site: http://www.ari.de/

TecPilot Rating: ✦ ✦ ✦ ✦

Compatible with: FS98

Requirements: Pentium 500, 64 MByte RAM, 3D Graphics Card.

Description: AriFlight's Amsterdam is small upgrade, which adds several buildings to the default flight sim scenery area. You will hardly notice the difference. In our view this type of scenery is BAD value for money.

In A Word: How can you sell products like this?

Title: ## Amsterdam Schiphol' 99

Publisher/Developer: Skysoft / Holger Schmidt

Web Site: http://www.simmarket.com/online

TecPilot Rating: ✦ ✦ ✦ ✦ ✦

Compatible with: FS98, FS2000

Requirements: Pentium 500, 64 MByte RAM, 3D Graphic Card

Description: A lot of 3D objects and buildings are added. Taxiway signs, Illuminated Touchdown Markers, Illuminated buildings, static ground vehicles etc.. The FS 98 version 1.0 can be upgraded via our homepage up to Ver1.4 for Free. The FS2000 Version already includes all updates.

In A Word: Done and done again!

Title: ## Amsterdam 2000

Publisher/Developer: Mice / Hans-Jörg Müller

Web Site: http://www.mice.ch/

TecPilot Rating: ✦ ✦ ✦ ✦

Compatible with: FS2000

Requirements: Pentium 500, 64 MByte RAM, 3D Graphic Card

Description: Includes: Runways and navaids - Roads - All buildings - Gate facilities of all terminals - Night lightning of the taxilines and taxiways - Night lightning Apron - Static aircraft at the piers - Docking systems -AGNIS/PAPA and SAFEGATE

In A Word: Oh, look! Another Schiphol Nice nonetheless!

Title: **Alicante 2001**

Publisher/Developer: Avion Magazine

Web Site: http://www.avionmagazine.com/alicante.htm

TecPilot Rating: ✦ ✦ ✦ ✦ ✦

Compatible with: FS2000

Requirements: Pentium 300, 128 MByte RAM, 3D Graphic Card

Description: Alicante, on the southern coast of Spain, features a very detailed airport, as well as surrounding towns, that include; photo-realistic terminal, hangars and service terminal buildings, and night lighting.

In A Word: A great addition for a great country.

Title: **Atlanta Intl Airport**

Publisher/Developer: SimFlyers

Web Site: http://www.simflyers.net/

TecPilot Rating: ✦ ✦ ✦ ✦ ✦

Compatible with: FS2000/FS2002

Requirements: Pentium II 300 PC, Windows 95/ 98, 64 Mb Ram, Sound Card

Description: SimFlyer's Atlanta scenery, is a true step forward for flight simulator sceneries. Aside from including detailed buildings, taxiways, and dynamic scenery, Atlanta sports a new feature, called serviceArmada. ServiceArmada creates realistic catering trucks, fuel trucks, and baggage carts, which appear at your airplane when you park, and dial in a certain frequency.

In A Word: A true innovation.

Title: **Bari Palese Airport**

Publisher/Developer: SimFlyers

Web Site: http://www.simflyers.net/

TecPilot Rating: ✦ ✦ ✦ ✦ ✦

Compatible with: FS2000/FS2002

Requirements: Pentium II 300 PC, Windows 95/ 98, 64 Mb Ram, Sound Card

Description: Bari Palese Airport is one of the most detailed Italian airports ever! It includes photorealistic buildings, night lighted taxiways, and buildings. The project also included FSTraffic tracks for the airport, and the surrounding area.

In A Word: A surefire classic.

SCENERY

Title: **Beijing China**

Publisher/Developer: SamSoft

Web Site: http://www.angelfire.com/fl/sammyscenery/kaitak.html

TecPilot Rating: ✦ ✦ ✦ ✦

Compatible with: FS98

Requirements: PC running Windows 95, 98, 2000 or NT, Pentium 200Mhz, 32MB of Ram,

Description: Beijing, capital of China, is an beautiful ancient city. This scenery includes Beijing's airport and the city view. They are all made according to the real world. Includes Airport , Summer Palace, Beijing Museum, Tiananman, many buildings, some are custom objects, different city textures, Olympic Stadium, Roads and Highways and the Beijing TV Tower.

In A Word: Nice but a little "old-hat" for flight sim these days

Title: **Budapest Scenery Millennium Edition**

Publisher/Developer: Andras Komez

Web Site: http://www.vistamaresoft.com/

TecPilot Rating: ✦ ✦ ✦ ✦

Compatible with: FS2000

Requirements: Pentium 600, 256 MByte RAM, 3D Graphics Card

Description: This extremely hi-resolution scenery of Budapest, come from Andras Kozma, one of flight simulations most respected scenery designers. Included is the city of Budapest (which includes thousands of buildings), and the region's five airports. The airports include detailed control towers, terminals, and taxiways.

In A Word: A solo project from one of the best!

Title: **Channel Islands**

Publisher/Developer: UK2000 Scenery / Gary Summons

Web Site: http://www.uk2000scenery.com

TecPilot Rating: ✦ ✦ ✦ ✦

Compatible with: FS2000, FS98

Requirements: Pentium II 300 PC, Windows 95/ 98, 64 Mb Ram, Sound Card

Description: Includes all the major Islands in full terrain mesh detail. With the airports of Jersey, Guernsey and Alderney. Even though this is a small scenery package it packs a big punch and generates loads of detail for an area that has always come last in FS.

In A Word: Setting standards

Title: **Chek Lap Kok**

Publisher/Developer: AriFlight

Web Site: http://www.ari.de/

TecPilot Rating: ✦ ✦ ✦ ✦ ✦

Compatible with: FS98

Requirements: Pentium 500, 64 MByte RAM, 3D Graphics Card.

Description: This is the depicts the all new Hong Kong Airport. It features most ground items including taxiways and photo realistic buildings and gates. More than 50 km of taxiways and center line lightings as well plus an authentic docking systems (Safe Gate). The package contains maps, the entire Cathay Pacific fleet as well as adventures for the self-appointed pilot.

In A Word: Wake up, we're here! - I said WAKE UP!

Title: **Colorado Rocky Mountains**

Publisher/Developer: Safari Fliteware

Web Site: http://www.fliteware.com/

TecPilot Rating: ✦ ✦ ✦ ✦ ✦

Compatible with: FS98, FS2000

Requirements: Pentium 500, 64 MByte RAM, 3D Graphics Card.

Description: Featuring well over 100 airfields from the eastern foothills to the western Colorado state border, Colorado is portrayed in custom 3D mesh terrain with thousands of discrete elevation points all taken from the USGS database. Hundreds of lakes and rivers are accurately portrayed as well as the standard Microsoft roads and highways.

In A Word: A little boring but great for the rednecks!

Title: **Courchevel Airport**

Publisher/Developer: The Reiffer Brothers

Web Site: http://www.reiffer.de/

TecPilot Rating: ✦ ✦ ✦ ✦ ✦

Compatible with: FS2000

Requirements: Pentium 500, 64 MByte RAM, 3D Graphics Card.

Description: Courchevel Airport is one of the most unusual airports The sloped runway and the surrounding mountains make every landing a real challenge. This 'altiport' is located in the French Alps close to Mont Blanc and Grenoble. These unique airport properties were recreated as realistically as possible within the limits of Microsoft Flight Simulator 2000. In addition to that this scenery makes use of the new FS2000 scenery code including night lighting, hard surface mesh terrain, maximum colours, animated objects and a fully functional seasonal texture change even for buildings.

In A Word: Close your eyes and think of England!

SCENERY

Title: | **El Hierro Scenery**

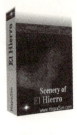

Publisher/Developer: HispaSim

Web Site: http://www.hispasim.com/

TecPilot Rating: ✦ ✦ ✦ ✦ ✦

Compatible with: FS2000

Requirements: Pentium 600, 256 MByte RAM, 3D Graphics Card

Description: El Hierro is the smallest island in the Canary chain. The scenery, created by HispaSim, allows you to fly into the island's airports, which include new buildings, and custom designed textures.

In A Word: A nice addition for the African flight junkies.

Title: | **English Airports**

Publisher/Developer: Barry Perfect

Web Site: http://www.gbairports.co.uk/

TecPilot Rating: ✦ ✦ ✦ ✦ ✦

Compatible with: FS98

Requirements: Pentium II 300 PC, Windows 95/ 98, 64 Mb Ram, Sound Card

Description: English Airports is a collection of airports including Heathrow, Gatwick, Stansted, Birmingham. Manchester and Luton. This collection is also available for FS2000 in boxed version published by Just Flight

In A Word: Created by an Englishman—looks good!

Title: | **Escalante and Hurricane**

Publisher/Developer: Georender

Web Site: http://www.georender.com/

TecPilot Rating: ✦ ✦ ✦ ✦ ✦

Compatible with: FS2000

Requirements: Pentium 600, 64 Mb RAM, 3D Graphics Card

Description: If you've never gone flying over Southern Utah, Escalante and Hurricane, will show you what its like. From the makers of the very successful freeware scenery, Flying M, Escalante and Hurricane include hi-resolution photorealstic airports, and scenery coving the 166 km that are between the two airports. They also include the surrounding towns, in amazing detail.

In A Word: Stirring up a hurricane of buzz!

Title:

European Airports 1 & 2

Publisher/Developer: AriFlight

Web Site: http://www.ari.de/

TecPilot Rating: ✦ ✦ ✦ ✦ ✦

Compatible with: FS98, FS95

Requirements: Pentium 500, 64 MByte RAM, 3D Graphic Card (Viper770 or similar)

Description: 13 airports situated all over Europe: from Gothenburg in the north towards Hanover, Bremen, Cologne/Bonn, to Luxembourg, the heart of Europe, from Lyon (Bron and Satolas), Milan (Malpensa and Linate) to Thessaloniki in the south.

In A Word: Terrible marketing, confusing package, bad all round! - Avoid!

Title:

Euro Collection

Publisher/Developer: Skysoft / Holger Schmidt

Web Site: http://www.simmarket.com/online

TecPilot Rating: ✦ ✦ ✦ ✦ ✦

Compatible with: FS98

Requirements: Pentium 500, 64 MByte RAM, 3D Graphic Card (Viper770 or similar)

Description: Airports included: Amsterdam, Athens, Barcelona, Duesseldorf, Geneva, Glasgow, Hamburg, Helsinki, London, Madrid, Manchester, Palma, Rome, Shannon, Vienna . Basic airport enhancement that's now catered for in FS2000.

In A Word: Very generic looking, but work well in FS98.

Title:

Fernando De Noronha & Rocas Alto

Publisher/Developer: Marcio N. Amaral

Web Site: http://www.realflight.com.br/

TecPilot Rating: ✦ ✦ ✦ ✦ ✦

Compatible with: FS2000

Requirements: Pentium 600, 128 Mb RAM, 3D Graphic Card

Description: The Fernando De Noronha and Rocas Alto depicts the two islands off of Brazil in stunning detail. The islands feature satellite textures, as well as many detailed landmarks, and airports. Static and dynamic scenery have also been included. The Fernando de Noronha was, in 1946, made into the Headquarters for the Brazilian Air force.

In A Word: Amazing!

Title: **Flight Adventure Asia**

Publisher/Developer: AriFlight

Web Site: http://www.ari.de/

TecPilot Rating: ✦ ✦ ✦ ✦ ✦

Compatible with: FS98

Requirements: Pentium 500, 64 MByte RAM, 3D Graphics Card

Description: This scenery boasts a load of features but in a nut-shell adds several buildings and airports through out all Asia. Basically a round up AriFlight's scenery packages featured in this book.

In A Word: Not worth it—Bad value for money.

Title: **FScene 2002**

Publisher/Developer: Ruud Faber

Web Site: http://www.fssharecenter.com/

TecPilot Rating: ✦ ✦ ✦ ✦ ✦

Compatible with: FS2000

Requirements: Pentium 600, 64Mb RAM, 3D Graphics Card

Description: This package included several re-done "3d looking" ground textures for Europe. The textures are said not to upset with FS2000 frame rates, or mesh terrain.

In A Word: No more US textures in Europe!

Title: **GB Airports 2001**

Publisher/Developer: Barry Perfect

Web Site: http://www.gbairports.co.uk

TecPilot Rating: ✦ ✦ ✦ ✦

Compatible with: FS2000

Requirements: Pentium 500, 64 Mb RAM, 3D Graphics Card

Description: GB Airports, is a continuation of English Airports for FS2000. GB include six of Great Britain's busiest airports, including; Glasgow, Edinburgh, Aberdeen, Newcastle, East Midlands and London City, all of which include true-to-life buildings, taxiways, and even wind socks.

In A Word: A sequel that is as good as the original!

Title: **German Airports Singles**

Publisher/Developer: Thomas Hirsch & Peter Hiermeie

Web Site: http://home.t-online.de/home/Th-Hirsch/Gap-Main.htm

TecPilot Rating: ✦ ✦ ✦ ✦ ✦

Compatible with: FS2000

Requirements: Pentium 350 MHz ore better. 64 MB RAM, better 128 MB. 2,8 MB free HD space. 3D graphics-accelerator card.

Description: The single airports of the successful GAP - German Airports series ! Containing Frankfurt, Cologne - Bonn, Leipzig + Hannover, Berlin Tegel, Düsseldorf, Hamburg + Bremen   this set of airports really show off FS2000's scenery engine.

In A Word: Say no more!

Title: **Gothenburg 2001**

Publisher/Developer: FS Dream Factory

Web Site: http://www.fsdreamfactory.com/

TecPilot Rating: ✦ ✦ ✦ ✦ ✦

Compatible with: FS2000

Requirements: Pentium 500, 64 Mb RAM, 3D Graphic Card

Description: This scenery package, from the FS Dream Factory, includes a highly detailed rendering of the city of Gothenburg, as well as the city's airport, Save. Al throughout the city, lay static cars, hi-resolution buildings, static ships in the ports, as well as other important land marks. The airport included accurate buildings, and taxi ways.

In A Word: Great Airport, amazing city!

Title: **Great Salt Lake**

Publisher/Developer: Safari Fliteware

Web Site: http://www.fliteware.com/

TecPilot Rating: ✦ ✦ ✦ ✦ ✦

Compatible with: FS2000

Requirements: Pentium 500, 64 MByte RAM, 3D Graphics Card.

Description: Encompassing 15,000 sq. miles, this scenery captures the diverse splendour of the Greater Salt Lake area at it's best. From the majestic mountains and magnificent waters to the Great Salt Desert, this full mesh 3D rendered scenery, provides you with magnificent detail for VFR flight at its finest.

In A Word: Yawn—but look at that scenery!

SCENERY

Title: **Himalaja**

Publisher/Developer: AriFlight

Web Site: http://www.ari.de/

TecPilot Rating: ✦ ✦ ✦ ✦ ✦

Compatible with: FS98

Requirements: Pentium 500, 64 MByte RAM, 3D Graphics Card.

Description: Including the whole area of Nepal, all airports with special structures, Katmandu International airport, near to the highest mountains in the world. Additionally Tibet China, parts of Indien, Pakistan and Bhutan ! 34 airports in total in the mountains - some of them are very dangerous ! See Mount Everest or K2 or look for the Yeti. You will be surprised.

In A Word: Very old but fun!

Title: **Hong Kong**

Publisher/Developer: Skysoft / Holger Schmidt

Web Site: http://www.simmarket.com/online

TecPilot Rating: ✦ ✦ ✦ ✦ ✦

Compatible with: FS95, FS98, FS2000 (Lite)

Requirements: Pentium 500, 64 MByte RAM, 3D Graphics Card.

Description: All Taxiway lines and RWY lights eg. Touchdown markers (night illuminated). All Taxiway signs (night illuminated). 2 Docking Gates. Static Planes. All buildings constructed or opened in Summer 1999. ILS and NAV's for RWY 07R/25L (RWY07L/25R was/is under construction). Night illuminated buildings. Incl an Airport Chart showing the Airport under construction.

In A Word: Old and past it, but great in its time—if you live in Hong Kong that is.

Title: **Ibiza 2001 v2**

Publisher/Developer: Sim-Wings FS Software

Web Site: http://www.simmarket.com/online

TecPilot Rating: ✦ ✦ ✦ ✦ ✦

Compatible with: FS2000

Requirements: Pentium 500, 64 MByte RAM, 3D Graphics Card.

Description: The scenery covers the airport of Ibizia. It replaces default scenery with photorealistic buildings, and night lighting effects.

In A Word: This guy is gonna be big!

Title: **Kai Tak (vol2)** (& surrounding)

Publisher/Developer: Samsoft

Web Site: http://www.samsoft-air.net/

TecPilot Rating: ✦ ✦ ✦ ✦ ✧

Compatible with: FS98

Requirements: PC running Windows 95, 98, 2000 or NT, Pentium 200Mhz, 32MB of Ram,

Description: Includes 100% realistic terminal buildings ! Advertising signs ! e.g. Coca Cola, Salem, Films, 555 tobacco, soyamilk, etc. Custom buildings ! e.g. blue buildings, aviation club, custom buildings around. Fences, airport buses, cargo vehicles, highway and taxi light Very beautiful night texture effects!

In A Word: Hong Kong here we come!

Title: **Kuala Lumpur Sepang 2000**

Publisher/Developer: Mice / Hans-Jörg Müller

Web Site: http://www.mice.ch/

TecPilot Rating: ✦ ✦ ✦ ✦ ✧

Compatible with: FS2000

Requirements: Pentium 500, 64 MByte RAM, 3D Graphics Card.

Description: Includes: Runways and navaids - Roads - All buildings - Gate facilities of all terminals - Nightlightning of the taxilines and taxiways - Nightlightning Apron - Static aircraft at the piers - Docking systems -AGNIS/PAPA and SAFEGATE - Dynamic Scenery with People Mover, 24 aircraft and 10 catering vehicles.

In A Word: Detailed but sparse in other areas.

Title: **Landvetter 2001**

Publisher/Developer: Cornel Grigoriu

Web Site: http://www.fsdreamfactory.com/

TecPilot Rating: ✦ ✦ ✦ ✧

Compatible with: FS2000

Requirements: Pentium 350, 64 Mb RAM, 3D Graphics Card

Description: Landvetter brings Sweden's second largest airport, Gothenburg Landvetter, to life in FS2000 with photo-realistic textures, and accurate taxiways. Landvetter is part of an ongoing project, called Gothenburg.

In A Word: A great start to an exiting project

SCENERY

Title: **Las Vegas**

Publisher/Developer: Safari Fliteware

Web Site: http://www.fliteware.com/

TecPilot Rating: ✦ ✦ ✧ ✧ ✧

Compatible with: FS2000

Requirements: Pentium 500, 64 MByte RAM, 3D Graphics Card.

Description: Updates casinos with enhanced night lighting, donated by Vincent Coene. The scenery extends from the Mountains West of Las Vegas, through Lake Mead and into the Western Grand Canyon, connecting it to Wilco Publishing Grand Canyon Scenery. You can follow the Colorado river from Laughlin, through treacherous canyons and over beautiful lakes. Encompassing over 15,000 sq. miles, this scenery captures the beauty of the Las Vegas area.

In A Word: Many might ask, "What's the point?" - Don't listen!

Title: **London Gatwick v2**

Publisher/Developer: Gary Summons

Web Site: http://www.uk2000scenery.com

TecPilot Rating: ✦ ✦ ✦ ✧

Compatible with: FS2000, FS98, FS2002

Requirements: Pentium II 300 PC, Windows 95/ 98, 64 Mb Ram, Sound Card

Description: London Gatwick is the UKs Second busiest airport and is also Londons main charter/IT airport, as with most UK Airports it is always being updated and extension added. This version includes the new front extension to the south terminal and the slight paint change to the North term.

In A Word: We did a double take on the price! - great stuff!

Title: **London Airports**

Publisher/Developer: Gary Summons / UK2000 Scenery

Web Site: http://www.uk2000scenery.com/

TecPilot Rating: ✦ ✦ ✦ ✧

Compatible with: FS2000, FS2002

Requirements: Pentium 600, 128 Mb RAM, 3D Graphics Card

Description: Gary Summons, the legendary airport designer, has lived up to his name in UK2000 Scenery - London Airports. The 5 airports (Heathrow, Stansted, Gatwick, Luton and London City), includes photo realistic terminals, ground textures, as well as dynamic scenery. Gary has also made British runway markings for the airports. Moving catering trucks are also found at most of the airports!

In A Word: The perfect cup of tea and works on lower end systems.

Title: **Madrid 2000**

Publisher/Developer: Avion Magazine

Web Site: http://www.avionmagazine.com/madrid.htm

TecPilot Rating: ✦ ✦ ✦ ✦ ✦

Compatible with: FS2000

Requirements: Pentium 350, 128 Mb RAM, 3D Graphics Card

Description: Part of the Spanish Global Scenery Project, Madrid 2000 features the Spanish capital's airport, as well as several dozen surrounding villages and towns. All of them are decorated with photorealistic buildings, and textures.

In A Word: A capital project!

Title: **Menorca 2001**

Publisher/Developer: Sim-Wings

Web Site: http://www.sim-wings.de/

TecPilot Rating: ✦ ✦ ✦ ✦ ✦

Compatible with: FS2000

Requirements: Pentium 500, 64 Mb RAM, 3D Graphics Card

Description: Menorca 2001 features a highly detailed airport, as well as dynamic and static aircraft. Also included is the "Aero club de Mahon," and the city of Ciutadella which included hotels, and other buildings

In A Word: Defining detail!

Title: **Orlando International Airport**

Publisher/Developer: SimFlyers

Web Site: http://www.simflyers.net/

TecPilot Rating: ✦ ✦ ✦ ✦ ✦

Compatible with: FS2000

Requirements: Pentium 350, 64 Mb RAM, 3D Graphics Card

Description: SimFlyer's rendition of Orlando International includes accurate buildings, taxiways, and other airport features. The airports "active jetway" feature not only docks with your plane, but it adjusts for it's height.

In A Word: Nice detailing

SCENERY

Title: **Pamplona 2001**

Publisher/Developer: Sim-Wings

Web Site: http://www.sim-wings.de/

TecPilot Rating: ✦ ✦ ✦ ✧ ✧

Compatible with: FS2000

Requirements: Pentium 500, 64 Mb RAM, 3D Graphics Card

Description: Pamplona is a beautiful rendition of this popular Spanish city. The add-on included Pamplona's airport, as well as the town, and the arena. Several other landmarks have been included. The buildings are all covered in high resolution textures.

In A Word: An $11 honeymoon!

Title: **Pennsylvania New Jersey**

Publisher/Developer: FSGenesis, Directflight / Horizon LLC

Web Site: http://www.fsaddon.com/

TecPilot Rating: ✦ ✦ ✦ ✦ ✦

Compatible with: FS2000

Requirements: PIII600, 128 Mb RAM, 16 Mb Riva TNT type configuration

Description: The first files are now available for download in the form of a hi-resolution mesh terrain update. Fully-customisable visibility limits, touchdown surface throughout, and at a detailed 150-meter elevation-point resolution. Roads, rivers, streams, enhanced cities, stadiums, prominent landmarks. Major hubs, airports - Pittsburgh International, Philadelphia International, Newark International. Regional Airports include Harrisburg International, Stewart International.

In A Word: Fine, if you're into that sort of thing.

Title: **Peru North and Lima**

Publisher/Developer: Alfredo Mendiola Loyola

Web Site: http://www.geocities.com/peruscenery/
PSEnglish.html

TecPilot Rating: ✦ ✦ ✦ ✧ ✧

Compatible with: FS2000

Requirements: Pentium 600, 128 Mb RAM, 3D Graphics Card

Description: Peru North and Lima covers exactly what the title suggests. The add-on inserts more then 64 new airports and aero domes into the simulator. All of the airports now have accurate runways and taxiways. In addition, the 99 square kilometer area around each airport included detailed topography, and cities.

In A Word: Lots of runway, little price.

Title: **Philadelphia Intl Airport**

Publisher/Developer: SimFlyers

Web Site: http://www.simflyers.net/

TecPilot Rating: ✦ ✦ ✦ ✦ ✦

Compatible with: FS2000

Requirements: Pentium 350, 64 Mb RAM, 3D Graphics Card

Description: The SimFlyer's PHL is one of the first ever payware renditions of the airport. The scenery features photo-realistic textures, as well as night lighting for taxiways, and buildings. Dynamic scenery is also included.

In A Word: Where's the crackers!

Title: **Polynesia 2000**

Publisher/Developer: Flight World

Web Site: http://www.flightworld.org/~Polynesia2000/

TecPilot Rating: ✦ ✦ ✦ ✦

Compatible with: FS2000

Requirements: Pentium 350, 128 Mb RAM, 3D Graphic Card

Description: Have you ever wanted to take a flight over French Polynesia? Flightworld's long awaited scenery add-on will let you. The islands are made of mesh terrain, and textured with high resolution satellite textures. Hotel's, churches, and buildings are all include

In A Word: The picture doesn't do it justice! Worth every penny! Only thing you don't get is a tan!

SCENERY

Title: **San Diego Scenery 2000**

Publisher/Developer: Safari Fliteware

Web Site: http://www.fliteware.com/

TecPilot Rating: ✦ ✦

Compatible with: FS98, FS2000

Requirements: Pentium 500, 64 MByte RAM, 3D Graphics Card.

Description: Aerial photos have been used to give extremely accurate color and to obtain a max resolution of 5.5 meters per pixel in detailed all-mesh terrain. It gives a photographic view of San Diego remarkably similar to that seen during actual flight. Roads, parks, and other landmarks are easily identified and those living in San Diego will be able to find their own house!

In A Word: Turning the corner in flight simulation scenery design!

Title: **San Francisco**

Publisher/Developer: Safari Fliteware

Web Site: http://www.fliteware.com/

TecPilot Rating: ✦ ✦ ✦ ✦ ✦

Compatible with: FS2000

Requirements: Pentium 500, 64 MByte RAM, 3D Graphics Card.

Description: Encompassing over 15,000 sq. miles, this scenery captures the beauty of the San Francisco Bay area, including Sacramento, Stockton and San Jose. This fully mesh scenery provides you with thousands of discrete elevation points and magnificent detail for VFR/IFR flight at its finest. The primary focus here is on terrain and all graphics are done in 16 bit High Colour.

In A Word: Boring—Strange airports— Great for cross country though!

Title: **Santander 2001**

Publisher/Developer: Sim-Wings

Web Site: http://www.sim-wings.de/

TecPilot Rating: ✦ ✦ ✦ ✦ ✦

Compatible with: FS2000

Requirements: Pentium 600, 128 Mb RAM, 3D Graphics Card

Description: Santander Airport, northern part of spain in Cantabria, now has a simulated counterpart! The Satander for FS2000 features hi-res mesh terrain, as well as airport buildings, and hangers.

In A Word: A nice upgrade for frequent flyers.

Title: **Sevilla**

Publisher/Developer: Avion Magazine

Web Site: http://www.avionmagazine.com/

TecPilot Rating: ✦ ✦ ✦ ✦ ✦

Compatible with: FS2000

Requirements: Pentium 350, 128 Mb RAM, 3D Graphic Card

Description: Sevilla is part of the Spanish Global Scenery Project. It includes a complete overhaul of Sevilla, San Pablo Intl., Cadiz, Cordoba, and La Juliana Aerodromes airports. Surrounding towns and villages have also been included.

In A Word: Excellent!

Title: **Sierra Nevada's 2000**

Publisher/Developer: Safari Fliteware

Web Site: http://www.fliteware.com/

TecPilot Rating: ✦ ✦ ✧ ✧ ✧

Compatible with: FS2000

Requirements: Pentium 500, 64 MByte RAM, 3D Graphics Card.

Description: Including the Yosemite National Park, Reno Nevada, Lake Tahoe, Crater Lake and east to the desert, including thousands of lakes, 50 airfields and several 3D objects, this fully mesh scenery provides you with thousands of discrete elevation points and magnificent detail for VFR/IFR flight at its finest.

In A Word: Lush Lush Lush!

Title: **Singapur**

Publisher/Developer: AriFlight

Web Site: http://www.ari.de/

TecPilot Rating: ✦ ✦ ✦ ✧ ✧

Compatible with: FS98

Requirements: Pentium 500, 64 MByte RAM, 3D Graphics Card.

Description: Exact area of Singapure and the south of Malaysia. More than 7000 km² flightarea. Detailed to 10 m / pixel. More than 400 buildings and objects. Dynamic landscapes. Vehicles and ships in movement and even a full functional tram. Includes Changi International.

In A Word: Small, but a real improvement

Title: **South China 2000**

Publisher/Developer: Samsoft—Sam Chan

Web Site: http://www.samsoft-air.net/

TecPilot Rating: ✦ ✦ ✦ ✧ ✧

Compatible with: FS98

Requirements: PC running Windows 95, 98, 2000 or NT, Pentium 200Mhz, 32MB of Ram,

Description: This scenery covers several of China's airports. The airports all include "100% accurate" taxiways, as well as new buildings and textures.

In A Word: Great designer, not so great scenery.

SCENERY

Title: **Sri Lanka - Special Edition**

Publisher/Developer: AriFlight

Web Site: http://www.ari.de/

TecPilot Rating: ✦ ✦ ✧ ✧ ✦

Compatible with: FS98

Requirements: Pentium 500, 64 MByte RAM, 3D Graphics Card.

Description: This scenery covers all of Sri Lanka with photo realistic satellite textures. All the textures are in 26 m/pixel resolution.

In A Word: One of the pioneers of photo realistic scenery.

Title: **South West England**

Publisher/Developer: UK2000 Scenery / Gary Summons

Web Site: http://www.uk2000scenery.com

TecPilot Rating: ✦ ✦ ✦ ✧ ✦

Compatible with: FS2000, FS98

Requirements: Pentium II 300 PC, Windows 95/ 98, 64 Mb Ram, Sound Card

Description: UK2000 Part2 is a collection of 25 Airports in the southwest of England made to a good quality. All the airports are based on real photograghs and aviation charts which sets a new standard in scenery design and brings the flight sim pilot closer to reality. FS98 Version Compatible with most UK scenery packages including Europe 3, Great Britain, England and Wales and VFR scenery.

In A Word: Just as good as the first!

Title: **South East England**

Publisher/Developer: UK2000 Scenery / Gary Summons

Web Site: http://www.uk2000scenery.com

TecPilot Rating: ✦ ✦ ✦ ✧ ✦

Compatible with: FS2000, FS98

Requirements: Pentium II 300 PC, Windows 95/ 98, 64 Mb Ram, Sound Card

Description: A collection of 24 Airports (22 in FS98 version) in the southeast of England made to remarkable quality and realism. All the airports are based on real photographs and aviation charts which sets a new standard in scenery design and brings the flight sim pilot closer to reality. This product also includes the London Gatwick product. (featured on the next page)

In A Word: Brutally accurate!

Title: The Olympic City 2000

Publisher/Developer: Samsoft

Web Site: http://www.samsoft-air.net/

TecPilot Rating: ✦ ✦ ✦ ✦ ✦

Compatible with: FS2000

Requirements: Pentium 200Mhz - 32MB of Ram - 23MB free hard disk space

Description: Realistic airport terminal textures. Static aircrafts, trucks, buses at the airport. Added Sydney style houses and mansions. Over 200 buildings and houses. Ships in the harbour. Sydney harbour view. Opera House has night effect. Includes the Darling Harbour, Manly Beach, Bondi Beach. Olympic Stadium is also included. The Sydney Tower ! The Rocks City is made with a real photo !

In A Word: Cute and cuddly!

Title: USA Collection

Publisher/Developer: Skysoft / Holger Schmidt

Web Site: http://www.simmarket.com/online

TecPilot Rating: ✦ ✦ ✦ ✦ ✦

Compatible with: FS95, FS98

Requirements: Pentium 500, 64 MByte RAM, 3D Graphics Card.

Description: Airports included: Atlanta, Boston, Chicago, Cincinnati, Denver, Houston, Las Vegas, Los Angeles, Miami, Minneapolis, Newark, NY Kennedy, NY La Guardia, Oklahoma, Pittsburgh, Salt Lake City, San Francisco, Seattle, St. Louis Intl, Washington DC. Generic looking but fine if you are used to FS98.

In A Word: Again, past it these days but great if you are stuck like glue to FS98.

Title: Vienna

Publisher/Developer: AriFlight

Web Site: http://www.ari.de/

TecPilot Rating: ✦ ✦ ✦ ✦ ✦

Compatible with: FS98

Requirements: Pentium 500, 64 MByte RAM, 3D Graphics Card

Description: The complete landscape of Vienna and the surrounding area - more than 7.500 km² . Detailed to 10 m / pixel. More than 20 touristic attractions. Beyond it there are included all 19 bridges and more than 25 buildings in the scenerie of Vienna. Including the biggest 11 airports in an area of 2000 km². "Europe 1" compatible.

In A Word: Not to original.

SCENERY

Title: **Virgin Islands**

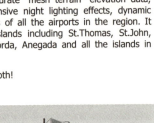

Anegada, B.V.I.

Publisher/Developer: Chris Wilkes

Web Site: http://www.simmarket.com

TecPilot Rating: ✦ ✦ ✦ ✦ ✦

Compatible with: FS2000

Requirements: Pentium 500, 64 MByte RAM, 3D Graphics Card.

Description: Features 6 metre per pixel resolution, accurate "mesh terrain" elevation data, hand colored and rendered bitmaps, extensive night lighting effects, dynamic elements and completely custom renditions of all the airports in the region. It covers all of the US and British Virgin Islands including St.Thomas, St.John, Tortola, St.Croix, Jost Van Dyke, Virgin Gorda, Anegada and all the islands in between.

In A Word: Popular—and rightly so! Calculated and smooth!

Title: **Waterworld 2000**

Publisher/Developer: Mice / Hans-Jörg Müller

Web Site: http://www.mice.ch/

TecPilot Rating: ✦ ✦ ✦ ✦ ✦

Compatible with: FS2000

Requirements: Pentium 500, 64 MByte RAM, 3D Graphic Card

Description: Imaginary scenery of Amsterdam "After the flood". Includes Runways and navaids - Gate facilities of all terminals - Windmills - Water, Water, Water !!!

In A Word: Where's my flippers!

SCENERY

Title: **White Wing Ranch**

Publisher/Developer: Marcio N. Amaral

Web Site: http://www.realflight.com.br/

TecPilot Rating: ✦ ✦ ✦ ✦ ✦

Compatible with: Pentium 350, 128 Mb RAM, 3D Graphic Card

Requirements: FS2000

Description: White Wing Ranch, in Brazil is a wonderful rendition of this tiny airport. Many textures have been created by hand. Winter, and Summer are also correctly displayed for a South American airport.

In A Word: Unique idea, and a great product!

Title: **Wonderful Rio**

Publisher/Developer:	Marcio N. Amaral
Web Site:	http://www.realflight-br.com/
TecPilot Rating:	✦ ✦ ✦ ✧ ✧
Compatible with:	FS2000

Requirements: Pentium 500, 64 MByte RAM, 3D Graphics Card

Description: Encapsulating the beautiful city if Rio de Janeiro, Marcio N. Amaral succeeds in re-creating the atmosphere of the stunning landscape in this complete package. To be tried and tested for sure!

In A Word: As the name implies.. Wonderful!

Title: **Zurich 2000**

Publisher/Developer:	Mice / Hans-Jörg Müller
Web Site:	http://www.mice.ch/
TecPilot Rating:	✦ ✦ ✦ ✦ ✦
Compatible with:	FS2000

Requirements: Pentium 500, 64 MByte RAM, 3D Graphics Card

Description: Zurich airport with terminal und fingerdock A,B, parking-, freightbuildings and hangars. Villages and towns located around the airport. Town of Winterthur, nearby villages and hills. Gate facilities of terminal A and B. Dynamic docking system at terminal A, gates A02, A08(only B737 and small planes) and A55. Dynamic docking system at terminal B, gates B31 and B33. Helicopter base. Operations center OPS, freight, GAC. Swissair hangars and loads more!

In A Word: Nice, but worried about frame rates!

Title: **Zurich 98**

Publisher/Developer:	Mice / Hans-Jörg Müller
Web Site:	http://www.mice.ch/
TecPilot Rating:	✦ ✦ ✦ ✦ ✧
Compatible with:	FS95, FS98

Requirements: Pentium 233, 64 MByte RAM, 3D Graphic Card

Description: Compatible with Apollo's Europe1. Zurich Airport with terminal and fingerdocks A,B, parking-, freightbuildings and hangars Villages and towns located around the airport. Town of Winterthur, nearby villages and hills. Gate facilities of terminal A and B Dynamic docking system at terminal A, gates A02, A08(only B737 and small planes) and A55. Dynamic docking system at terminal B, gates B31 and B33. Helicopter base. Scenery covered with snow in winter, landscape of Europe1 and Loads more!

In A Word: If you still have FS95 or F98 this is great!

SCENERY

LAST MINUTE ADDITIONS

Title:	**East Anglia & North London**
Publisher/Developer:	UK2000 Scenery / Gary Summons
Web Site:	http://www.uk2000scenery.com
TecPilot Rating:	✦ ✦ ✦ ✦ ✧
Compatible with:	FS2000, FS2002
Requirements:	Pentium 600, 64 MByte RAM, 3D Graphics Card.
Description:	Features 33 Airports and airfields in eastern England including London Heathrow and Stansted. Over 2500 purpose made textures were made for this project which over 2000 hours to make. Includes interactive objects, perimeter fencing, walls, trees, bushes, gates and docking systems.
In A Word:	One of the first releases for FS2002.

Title:	**Taipei, Taiwan**
Publisher/Developer:	Samsoft
Web Site:	http://www.samsoftpub.com
TecPilot Rating:	✦ ✦ ✦ ✧ ✧
Compatible with:	FS2000, FS2002
Requirements:	Pentium 600, 64 MByte RAM, 3D Graphics Card.
Description:	Realistic terminal buildings, Taxi lines, trees, Taiwan style houses around the airport, CKS airport and Sung Shan Airports, ships in the sea, Taipei "Memorial Hall", Grand Hotel and lots more
In A Word:	Dedicated designers - making the most out of what's available to them.

Title:	**The Ultimate Hong King**
Publisher/Developer:	FlightSoft
Web Site:	http://www.flightsoft.com
TecPilot Rating:	✦ ✦ ✦ ✧
Compatible with:	FS2000, FS2002
Requirements:	Pentium 600, 64 MByte RAM, 3D Graphics Card.
Description:	Includes 67 "sightseeing flighs" to explore the region. Details such as roadways, street signs, cars parked in garages, buses are clearly visible each placed carefully on an extremely detailed textured ground.
In A Word:	Too much detail to mention in this book - literally!

a-Z
Download Add-on Products
Utilities

Title: Aircraft Designer 2000

Publisher/Developer: Aeronautic Conceptions Team

Web Site: http://actpub.com/

TecPilot Rating: ✦ ✦ ✦ ✦ ✦

Compatible with: FS98, FS2000, CFS, CFS2

Requirements: Pentium 166 minimum (recommended), 32 Mb Ram, Soundblaster Compatible Sound Card, 2D/3D Card, Headphones or Speakers recommended.

Description: This program lets you make your very own planes for FS using wireframes. The program also allows you to add moving parts, and even paint your project.

In A Word: Extremely advanced, not for the faint hearted.

Title: Architect 2000

Publisher/Developer: Doug Pouk

Web Site: http://www.flightsimmers.net/airport/architect2/

TecPilot Rating: ✦ ✦ ✦ ✦ ✦

Compatible with: FS2000, FS2002

Requirements: Pentium 500, 64 MByte RAM, 3D Graphic Card

Description: Import default scenery from FS2000 scenery files. This feature makes it easier than ever to "spiff up" the existing Flight Simulator scenery. Seasonal and night textures. Create Airport Facility entries using the Architect Facility Creation Wizard. Architect 2000 airport entries will now appear in the Flight Simulator's Airport Facility, Flight Planner, and GPS display. Easily lay out photo-realistic ground textures using the Photo Texture Grid. Use aerial photographs to create scenery that is "As Real as it Gets!" Use Metric measurements. Automatically use FS2000 world magnetic variation settings. Easier polygon, road, line, taxiway point-by-point editing.

In A Word: Needs some perfecting, but there is potential.

Title: Blue Sky

Publisher/Developer: Iain Murray

Web Site: http://flightsim.computing.dundee.ac.uk/

TecPilot Rating: ✦ ✦ ✦ ✦ ✦

Compatible with: FS98,CFS

Requirements: Pentium 300 minimum, 64 Mb Ram.

Description: This utility replaces FS98 and CFS's purplish skies with a more realistic blue color.

In A Word: Fixes your pet-peeves

UTILITIES

Title: ### Combat Expansion Pack

Publisher/Developer: DirectFlight / Horizon LLC

Web Site: http://www.fsaddon.com/

TecPilot Rating: ✦ ✦ ✦ ✦ ✦

Compatible with: CFS

Requirements: Pentium 500, 64 MByte RAM, 3D Graphic Card

Description: CEP is a tool that puts all the options from FS98, into CFS. You can now use an auto-pilot, improve weather realism, and even create FS98 flight plans.

In A Word: Great idea!

Title: ### CoPilot v2.5

Publisher/Developer: Abacus

Web Site: http://www.abacuspub.com

TecPilot Rating: ✦ ✦ ✦ ✦ ✦

Compatible with: FS98, FS2000

Requirements: Microsoft Flight Simulator 98 Pentium II 300 PC, Windows 98, 98, 64 Mb Ram, CD ROM Drive, Sound Card.

Description: CoPilot is one of the most successful navigation tools for flight simulator, ever. The program is suited for all skill levels of navigators. CoPilot allows you to print out approach plates, ILS information, and much more!

In A Word: Obsolete nowadays

Title: ### CoPilot v2.8

Publisher/Developer: Abacus

Web Site: http://www.abacuspub.com

TecPilot Rating: ✦ ✦ ✦ ✦ ✦

Compatible with: FS2002

Requirements: Microsoft Flight Simulator 2000/98 Pentium II 300 PC, Windows 98, 98, 64 Mb Ram, CD ROM Drive, Sound Card.

Description: CoPilot has gained a reputation as the best selling flight sim utility. Its simple point-and-click flight planning, superior GPS navigation, detailed printed flight maps, instrument approach plates and airport charts are all features that users have asked for in this all-around first choice.

In A Word: Basic enough for the newbie, loaded enough for the vets.

UTILITIES

Title: ## Custom Panel Designer

Publisher/Developer: Abacus

Web Site: http://www.flightsimdownloads.com

TecPilot Rating: ✧ ✧ ✧ ✧ ✧

Compatible with: FS2000, CFS, FS98

Requirements: Pentium 200 or higher. Windows 95/98. 32mb RAM.

Description: Now anyone can make their own panels! Whether you're doing minor "touch up" work or designing a complex multi-window panel, you can count on Custom Panel Designer to make the work fast and easy.

In A Word: Gauges, gauges everywhere!

Title: ## EZ-GPS

Publisher/Developer: Abacus

Web Site: http://www.flightsimdownloads.com

TecPilot Rating: ✧ ✧ ✧ ✧ ✧

Compatible with: FS2000, CFS, FS98

Requirements: Pentium 200 or higher. Windows 95/98. 32mb RAM.

Description: A simple utility, created by Peter Jacobson that allows users of FS98 and CFS to easily navigate and fly direct between two points anywhere in the world.

In A Word: No Nonsense Navigation

Title: ## EZ-VFR

Publisher/Developer: Abacus

Web Site: http://www.abacuspub.com

TecPilot Rating: ✧ ✧ ✧ ✧ ✧

Compatible with: FS2000

Requirements: Pentium 600, 64 MByte RAM, 3D Graphic Card

Description: EZ-VFR is a small program that actually labels the airports you fly over in real time. It will display all airports with a 30 mile are of your position.

In A Word: To bad they don't have this in the real world!

UTILITIES

Title: ### Flight Manager

Publisher/Developer: Josef Dirnberger

Web Site: http://www.fssharecenter.com

TecPilot Rating: ✦ ✦ ✦ ✦ ✦

Compatible with: FS2000, FS98, CFS

Requirements: Pentium 200 with 200 Mhz, 32 MB of RAM, 20 MB of available Hard disk space, Flight Simulator 2000, 98, 95 or Combat Flight Simulator.

Description: Flight Manager is a new flight planner loaded with features. You can create your own SID and STAR charts as well as export your files to Radar Contact, or SwarkBox.

In A Word: Very realistic program.

Title: ### FS Mesh

Publisher/Developer: Burkhard Renk

Web Site: http://www.fssharecenter.com

TecPilot Rating: ✦ ✦ ✦ ✦ ✦

Compatible with: FS2000, F98

Requirements: PC CD-ROM. IBM PC & compatibles. Pentium 200 or higher. Windows 95/98. 32mb RAM. SVGA graphics card required.

Description: Intended to integrate the creation of big, high detail elevated mesh terrain and base scenery objects. You can modify every input point to correct for errors or to adjust to your sceneries need. Input to FSMesh has not to be in any grid form and can generate mesh down to the highest of detail.

In A Word: Knitting the scenery nicely!

Title: ### Final Approach

Publisher/Developer: Georges Lorsche

Web Site: http://ourworld.compuserve.com/homepages/glorsche/

TecPilot Rating: ✦ ✦ ✦ ✦ ✦

Compatible with: FS2000, FS98, FS95

Requirements: PC CD-ROM. IBM PC & compatibles. Pentium 200 or higher. Windows 95/98. 32mb RAM. SVGA graphics card required.

Description: Final Approach is a powerful, Windows-based Approach Chart Atlas and Designer for Microsoft Flight Simulator and other realistic general aviation flight simulators. Over 4300 approach charts can be downloaded for use with the program.

In A Word: Mapping the world!

UTILITIES

Title: **FS Meteo**

Publisher/Developer: Marc Phillibert

Web Site: http://fsmeteo.com

TecPilot Rating: ✦ ✦ ✦ ✦ ✦

Compatible with: FS2002, FS2000, FS98 & CFS 2

Requirements: Pentium 300 minimum 64 Mb Ram, Sound Card, 2D/3D Card,

Description: This utility allows you to download real world weather from the net, and put it in flight simulator. Or, if you don't want to connect only, you can adjust wind layers and other weather features offline.

In A Word: Great, if your system can handle 8 different cloud layers, plus winds aloft etc.

Title: **FS Navigator**

Publisher/Developer: Helge Schroeder

Web Site: http://fsnavigator.com/

TecPilot Rating: ✦ ✦ ✦ ✦ ✦

Compatible with: FS98, FS2000, FS2002

Requirements: Pentium 300 minimum, 64 Mb Ram. (FS98) 128Mb Ram (FS2000/02)

Description: Integrated Module for FS98 and FS2000, FS2002 . Navigational Map and Flight Planner. Flight Management System. Automatically flies flight planned routes and holdings. World-wide Search & Find function for Airports and Navigational Aids. Additional Autopilot for comfortable operation. World-wide current Airway and Intersection Database also for FS98. Graphical SID/STAR Editor. SID/STAR direct Up-/Download from dedicated Internet Server. Automatic Routing by Airways and SID/STARs. Flight Simulator compatible Multiplayer client for Internet and Intranet with display all involved airplanes .. and lots more. Constantly updated.

In A Word: A little "all over the place" but VERY concise!

Title: **FS NavPak 2K™**

Publisher/Developer: DirectFlight / Horizon LLC

Web Site: http://www.fsaddon.com/

TecPilot Rating: ✦ ✦ ✦ ✦ ✦

Compatible with: FS98, FS2000

Requirements: Pentium 500, 64 MByte RAM, 3D Graphic Card

Description: Features include: VORs and VORTACs—specify the frequency, range, and even enable DME! ADF—adjust location (including elevation) and frequency! ILS—in addition to localizer position, orientation, and glideslope angle! Placing marker beacons has never been easier - simply specify a distance and the rest is automatic! Placing runways has never been easier! Runway characteristics—adjust dimensions, type, and markings!

In A Word: A fix that lets you make your own fixes.

Title: FSPlanner 4.0

Publisher/Developer: Sascha Felix

Web Site: http://www.phil.uni-passau.de/linguistik/fsp

TecPilot Rating: ✦ ✦ ✦ ✦ ✦

Compatible with: FS95, FS98

Requirements: Pentium 166, 32 Mb Ram, Sound Card, 2D/3D Graphics Card

Description: Create flight plans (either manually or automatically), calculate fuel, alternate airports, look at maps with flight-relevant information (navaids, airports, runways), etc. In addition to flight planning, the program provides you with a GPS, two Moving Maps and a Weather Generator which produces weather changes during a flight.

In A Word: There are better.

Title: FsXfer

Publisher/Developer: Gert Heijnis

Web Site: http://www.flywarenl.com/jetop/fsxfer.htm

TecPilot Rating: ✦ ✦ ✦ ✦ ✦

Compatible with: FS2000

Requirements: Pentium 600, 64 MByte RAM, 3D Graphic Card

Description: This package of files allows you to spread your view of FS2000 over several screens, via a network. You can also move the instrument panels to different screens.

In A Word: Spread your views and fly!

Title: FS Scenery Enhancer

Publisher/Developer: Lago

Web Site: http://www.lagoonline.com

TecPilot Rating: ✦ ✦ ✦ ✦ ✦

Compatible with: FS2002

Requirements: Pentium II 300, 64Mb RAM, HD space 650/670Mb, Audio Card, Video Card 8MB, 3D acceleration

Description: FS Scenery Enhancer is a unique product that makes it very simple to add scenery elements (objects) to existing scenery. What makes this product special is that the whole process of scenery enhancing takes place INSIDE Flight Simulator 2002 and not in a separate program as was the only option previously. There is no code to write no compiling needed. amazing

In A Word: A unique and completely novel concept

Before

After

UTILITIES

FS Scenery Manager

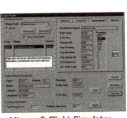

Publisher/Developer:	Michael Garbers
Web Site:	http://scenery-manager.isapage.com/
TecPilot Rating:	✦ ✦ ✦ ✦ ✦
Compatible with:	FS2000, FS2002 & CFS2
Requirements:	Pentium 300 minimum, 64 Mb Ram, Soundblaster Compatible Sound Card, 2D/3D Card.
Description:	This program makes it possible to manage and control original and custom scenery created for Flight Simulator (2000, 2002, CFS II). There is no need worry about the airport facilities directory as it will automatically read all installed sceneries into its own database. It also includes an automatic system to install scenery from a source path or even directly out of a ZIP-File.
In A Word:	Only if you are a scenery buff who has 300 add-ons to manipulate.

Jet Operator

Publisher/Developer:	Gert Heijnis
Web Site:	http://www.flywarenl.com/jetop
TecPilot Rating:	✦ ✦ ✦ ✦ ✦
Compatible with:	FS98, FS2000
Requirements:	Pentium 300 minimum (recommended), 64 Mb Ram, Microsoft Flight Simulator 98, Soundblaster Compatible Sound Card, 2D/3D Card, Headphones or Speakers recommended.
Description:	JetOp is a powerful tool that allows you to edit the flight dynamics of aircraft to suite your needs.
In A Word:	Only for TRUE enthusiasts.

MicroPanel Studio

Publisher/Developer:	Gert Heijnis
Web Site:	http://www.flywarenl.com/jetop
TecPilot Rating:	✦ ✦ ✦ ✦ ✦
Compatible with:	FS2000, FS2002
Requirements:	Pentium 300, 64 Mb Ram, Soundblaster Compatible Sound Card, 2D/3D Card.
Description:	MicroPanel Studio is a design time and run time environment for creating micropanels. A micropanel is a popup information display that can be shown or hidden when flying with Microsoft Flight Simulator. These micropanels can be whatever size, shape and color you wish. They can display any information you require in whatever format you need. There are over 300 different MS Flight Simulator variables available for use.
In A Word:	Only for TRUE enthusiasts.

Title: ## Moving Map

Publisher/Developer: John Hnidec

Web Site: http://www.simsystems.com.au/

TecPilot Rating:

Compatible with: FS95, FS98

Requirements: Pentium 166, 32 Mb Ram, Sound Card, 2D/3D Graphics Card

Description: This moving map displays state borders, Navaids, as well as airports, but it does not hurt the frame rate of the simulator.

In A Word: And on the 8th day, Microsoft gave us a GPS in FS2000.

Title: ## NOVA

Publisher/Developer: Rafael Sanchez

Web Site: http://www.fsnova.com/

TecPilot Rating:

Compatible with: FS2000, FS2002

Requirements: Pentium 600, 64 MByte RAM, 3D Graphic Card

Description: NOVA was one of the first utilities to design 3D objects for Microsoft's Flight Simulator. NOVA can help you build custom hangars, control towers, buildings, bridges, terminals, gates, houses, docking systems and much more.

In A Word: Your mind is the limit!

Title: ## QuickMap

Publisher/Developer: Project Magenta

Web Site: http://www.projectmagenta.com

TecPilot Rating:

Compatible with: FS2002, FS2000, FS98

Requirements: Pentium 600, 64 MByte RAM, 3D Graphic Card

Description: In a nut shell, this is basically a moving map similar to the one that comes with FS2000 and FS2002 with the addition of an ILS profile display, VNAV display, AI Aircraft display and dynamic repositioning of aircraft. Apart from a host of other features.

In A Word: Updates a pretty poor default GPS admirably.

Title: **Scenery Apprentice**

Publisher/Developer: Abacus/Louis Sinclair

Web Site: http://www.abacuspub.com

TecPilot Rating: ✦ ✦ ✦ ✦ ✦

Compatible with: FS2000

Requirements: Pentium 600, 64 MByte RAM, 3D Graphic Card

Description: Scenery Apprentice is a great program for those of us who want to touch up local scenery, but don't want to have to learn complicated scenery languages, and deal with complicated programs. The program lets you create objects, then place them around the airport.

In A Word: Making a difficult job easy

Title: **Toposcanner**

Publisher/Developer: Patrick Bernhart

Web Site: http://www.fssharecenter.com/

TecPilot Rating: ✦ ✦ ✦ ✦ ✦

Compatible with: FS2000

Requirements: Pentium 350, 64 MByte RAM, 3D Graphic Card

Description: Toposcanner is a program that allows the user to view, and strip latitude and longitude readings from Digital Elevation Data maps (DEMs), and use them in FS2000.

In A Word: One of a kind and extremely underestimated and understood!

Title: **Ultimate Airlines**

Publisher/Developer: Flight One

Web Site: http://www.flight1.com

TecPilot Rating: ✦ ✦ ✦ ✦ ✦

Compatible with: FS2000

Requirements: Pentium 600, 64 MByte RAM, 3D Graphic Card

Description: Ultimate Airlines will provide time the time tables for most all airline flights in the whole world! The program allows realistic flight plans to be created, and be exported to AETI's ProFlight, to be turned into voice ATC flights.

In A Word: The ultimate airline operations utility!

Title: **Visual Aircraft Studio**

Publisher/Developer: Jorge Santoro & Marco Silva

Web Site: http://www.task.com.br/users/santoro/vas.htm

TecPilot Rating: ✦ ✦ ✦ ✦ ✦

Compatible with: FS98

Requirements: Pentium 500, 64 MByte RAM, 3D Graphic Card

Description: Visual Aircraft Studio (VAS) is a tool created to make the aircraft painting process easier for both Microsoft Flight Simulator and Combat Flight Simulator. It brings a powerful three-dimensional interface that allows real time visualization of the aircraft being painted, to provide an immediate view of the the painted plane as it will be displayed in MS Flight Simulator. VAS provides unique features, which makes the aircraft painting process very easy and intuitive. NOTE: This Utility requires Flight Shop or Aircraft Factory 99

In A Word: How do get from "this" to "TH-H-ISS-S"

Title: **Voice Master Pro**

Publisher/Developer: Manuel Santos

Web Site: http://members.home.com/voice-master

TecPilot Rating: ✦ ✦ ✦ ✦ ✦

Compatible with: FS98, FS2000

Requirements: Pentium 500, 64 MByte RAM, 3D Graphic Card

Description: Voice Master enables the user to issue voice commands with the purpose of controlling the game environment. Additionally, provides a speech synthesis engine that simulates the voice of your co-pilot and can interact with you based on the commands you issue.

In A Word: Speak to FS with ease!

Title: **Wetter 2000**

Publisher/Developer: Klaus Prichatz

Web Site: http://fssharecenter.com/wetter

TecPilot Rating: ✦ ✦ ✦ ✦ ✦

Compatible with: FS2000, FS98

Requirements: Pentium 300 minimum (recommended), 64 Mb Ram, Soundblaster Compatible Sound Card, 2D/3D Card, Headphones or Speakers.

Description: Wetter 2000 gives you the power to create dynamic weather, and enroute weather conditions.

In A Word: Great for long haul flights.

UTILITIES

Reference

(ft/sec) = 0.6818 mph
(ft/sec) = 1.097 km/h
(hp) = 33000 ft.lb/min
(hp) = 745.7 watts
(kt) = 0.5151 m/s
(kt) = 1.151 mph
(kt) = 1.852 km/h
(mph) = 0.4471 m/s
(mph) = 0.8684 kt
(mph) = 1.609 km/h
Cubic foot (ft3) = 2.832 x 10-2 cubic metres
Cubic foot (ft3) = 28.317 litres
Cubic inch (m3) = 1.639 x 10-5 cubic metres
Feet (ft) = 0.3048 metres
Feet(ft) = 3.048 x 10-4 kilometres
Foot lb./sec (ft.lb/sec) = 1.356 Joules/s (watts)
Foot Pound (ft.lbf) = 1.356 Nm (Joules)

Foot/second (ft/sec) = 0.3048 m/s
Horsepower (hp) = 550 ft.lb/sec
Imp.Gallon (Imp gal) = 4.546 litres
Inch (in) = 25.40 millimetres
Knot [nm/hr] (kt) = 1.689 ft/sec
Mile/hour (mph) = 1.467 ft/sec
Nautical mile (nm) = 1.1508 statute miles
Nautical mile (nm) = 1.852 kilometres
Pound (lbf) = 4.448 N
Pound/Sq. ft (lbf/ft2) = 47.88 N/m2
Pound/Sq. in (lbf/in2) = 6895 N/m2
Slug (slug) = 14.59 kg
Slug/cubic ft (slug/ft3) = 515.4 kg/m3
Square foot (ft2) = 9.290 x 10-2 square metres
Statute mile(mi) = 1.609 kilometres
U.S.gallon (US gal) = 3.78542 litres

Derived Units

Coef.of Viscosity: Slug/ft sec= 47.88 kg/ms
Density: Slug/ft3 = 515.4 kg/m3 (1 lb/ft3 = 16.02 kg/m3)
Force: Pound (lbf) = 4.448 N
Gas Constant: ft lbf/slug°R = 0.1672 Nm/kg°K
Kinematic Viscosity: ft2/sec = 9.290 x 102 m2/s
Power: Slug ft2/sec3 = 1.356 Nm/s
Pressure: Slug/ft sec2 = 47.88 N/m2 (1 lb./in2 (psi) = 6895 N/m2)
Specific fuel consumption: (jet a/c) lb/hr. lb thrust = 0.283 x 10-4 kg/Ns
Work: Slug ft2/sec = 1.356 Nm

Constants

Gravitational Force at SL = 9.80665 m/s2 = 32.174 ft/sec2
Air pressure at SL po = 760 mm Hg = 29.92 m.Hg = 1.01325 x 105 N/m2 = 2116.22 lb./ft2
Air temperature at SL To = 15.0° C (conversions ß)

= 288.15° K (°K = °C + 273.19)
= 59.0° F° (°C = (°F - 32) (5/9))
= 518.67° R (°R = °F + 459.7)

Air density at SL ro = 1.22492 Kg/m3 = 0.002378 slug/ft3
Air coef.of viscosity SL mo = 1.7894 x 10-5 kg/ms = 1.2024 x 10-5 lb/ft s
Air Kinematic viscosity SL no = 1.4607 x 10-5 m2/s = 1.5723 x 10-4 ft2/s

Specific gravity at 0°C (lb/ft3 / kg/m3):

Water =1.000 (62.43 /1000) Sea water = 1.025 (63.99 / 1025)
Jet Fuel JP1 = 0.800 (49.9 / 800) JP3 = 0.775 (48.4 / 775)
JP4 = 0.785 (49.0 / 785) JP5 = 0.817 (51.0 / 817)
Kerosine = 0.820 (51.2 / 820) Gasoline = 0.720 (44.9 / 720)
Alcohol = 0.801 (50.0 / 801)

Nautical mile (nm): The international nautical mile is 1852m exactly. The British nautical mile is 6080 feet but aviators and other navigators have sometimes used 6000 ft (2000 yards) as a crude approximation.

Knot (kt): is (nm/hour) = 0.514 m/s = 1.852 km/hr = 1.688 ft/sec = 1.1508 mph.

G [gee] (g): is the gravitational acceleration at ground level (average)

= 32.2 ft/sec2 = 9.81 m/sec2. Gee is used to divide aircraft accelerations to relate them to steady flight conditions (i.e. the aircraft is pulling 6g).

Slug: is a contrived unit used in the old British system devised to avoid multiplying by gravitational acceleration. It is a mass unit which provides a force of 1 pound when it is subjected to an acceleration of 1 ft/sec2. It is often used for specifying air density in aerodynamic equations in ft:lb:sec units (e.g.air at SL.ISA, ro= 0.002378 slug/ft3 which is equivalent to 1.225 Kg/m3).

British Thermal Unit (BTU): The heat required to raise the temperature of one pound of water through 1°F. [Note, 1BTU = 1055 Joule]

Horsepower (hp): An artificial measure of power = 550 ft.lb/s =3300 ft lb/min. = 746 watts

Bar (bar): An measurement of pressure = 106 dyne/cm2 (often quoted in millibars = 10 kN/m2).

Imperial Gallon (Imp gal): The volume of 10 pounds of water at 62°F = 277.4 in3 = 4.546 litre.

Ton: A measure of weight = 2240 pounds = 1016 kg (f) (approx. 1000 kg.)

Thou: One thousandth of an inch = 0.001in. [1mm = 40 thou.approx.]

Flight Level: is a derived unit of altitude used in air traffic control. One hundred feet altitude is the basic unit therefore flight level 330 is in practice at a height of 33000 ft.

Drag Count: is used as a crude measure for the change in drag coefficient (CD) = 0.0001

Historic People

Celsius: Swedish astronomer, Ander Celsius (1701-1744)

Hertz: German physicist, H.R. Hertz (1857-1894)

Joule: English physicist, J.P. Joule (1818-1889)

Kelvin: Scottish physicist, Lord W.T. Kelvin (1827-1907)

Newton: English mathematician, Sir I. Newton (1642-1727)

Pascal: French Scientist, Blaise Pascal (1623-1662)

Rankin: Scottish engineer, W.G.M Rankin (1820-1872)

Reynolds: British engineer, Prof.Osborne Reynolds (1842-1912)

Watt: Scottish engineer, James Watt (1736-1819)

Museums

Museums in the United Kingdom

Name: Bournemouth Aviation Museum
Addr: Hangar 600, Bournemouth International Airport, Christchurch, Dorset, ENGLAND, BH23-6SE
Tel: +44 (0) 1202-580858
Web: http://www.aviation-museum.co.uk
Opening/Closing: 10.00 - 16.00

Name: British Balloon Museum
Addr: 75 Albany Road , Old Windsor , Berkshire SL4 2QD, England
Tel: +44 (0) 1753 862977
Web: http://www.reading.ac.uk/AcaDepts/sn/wsn1/dept/av/en29.html
Opening/Closing: Monday, Tuesday, Thursday & Saturday 10.00 - 18.00

Name: Brooklands Museum
Addr: Brooklands Rd, Weybridge, Surrey, England, KT13 0QN.
Tel: +44 (0) 1932 857381
Web: http://www.motor-software.co.uk/brooklands/index.html
Opening/Closing: N/A

Name: The City of Norwich Aviation Museum
Addr: Norwich Airport in the village of Horsham St. Faith.
Tel: N/A
Web: http://www.mth.uea.ac.uk/~h720/aviation/conam.html
Opening/Closing: N/A

Name: Eden Camp
Addr: Malton, North Yorkshire
Tel: +44 (0) 1653 697777
Web: http://www.edencamp.co.uk/
Opening/Closing: 10.00 am - 5.00 pm every day from 2nd Monday in January until 23rd December

Name: Fleet Air Arm Museum
Addr: Near Ilchester, Somerset, BA22 8HT
Tel: +44 (0) 1935 840565
Web: http://www.faam.org.uk/
Opening/Closing: Call for times.

Name: Jet Age Museum
Addr: Hanger 7, Meteor Business Park, Gloucestershire Airport, Cheltenham Road East, Gloucester, GL2 9QL
Tel: +44 (0) 1452-715100
Web: http://gac.future.easyspace.com/gac.htm
Opening/Closing: Daily including Saturdays & Sundays, Easter to October - 10 am to 4 pm

Name: Lincolnshire Aviation Heritage Centre
Addr: East Kirkby, Spilsby, Lincolnshire. PE23 4DE England
Tel: +44 (0) 1790 763207
Web: http://freespace.virgin.net/nick.tasker/ekirkby.htm
Opening/Closing: Easter - Oct: Mon-Sat 10.00-17.00—Nov- Easter : Mon - Sat 10.00-16.00

Name: Metheringham Visitor Centre
Addr: Westmoor Farm, Metheringham, Lincolnshire LN4 3BQ
Tel: +44 (0) 1526 378270
Web: http://www.oden.co.uk/jherbert/metheringham_airfield/
Opening/Closing: Easter—Oct: Wednesdays 12.00 noon - 4.00pm Bank Holidays 12.00 noon - 5.00pm

Name: Midland Air Museum
Addr: Coventry Airport, Baginton, Warwickshire, CV8 3AZ, United Kingdom.
Tel: +44 (0) 2476 301033 (+44 2476 301033 from outside the UK).
Web: http://www.jetman.dircon.co.uk/mam/index.htm
Opening/Closing: (April to October) from 10:00 AM to 5:00 PM (November to March) from 10:30 AM to 4:30

Name: Montrose Air Station Museum
Addr: Waldron Road, Montrose, Angus
Tel: +44 (0)1674 675401
Web: http://ourworld.compuserve.com/homepages/AirspeedNews/
Opening/Closing: Sundays; 12 - 5 p.m

Name: The Museum of Army Flying
Address: Middle Wallop, Stockbridge, Hampshire, SO20 8DY
Tel: +44 (0)1980 674421
Web: http://www.flying-museum.org.uk/
Opening/Closing: Daily 10:00am to 4:30pm

Name: The Museum of Science and Industry in Manchester
Addr: The Museum is in Castlefield, minutes from the City Centre
Tel: +44 (0)161 832 2244
Web: http://ourworld.compuserve.com/homepages/AirspeedNews/
Opening/Closing: Sundays; 12 - 5 p.m

Name: North Weald Airfield Museum Association
Addr: Ad Astra House, 6 Hurricane Way, North Weald, Essex. CM16 6AA
Tel: N/A
Web: http://fly.to/northweald
Opening/Closing: March until December, on Saturday and Sundays from noon till 5pm

Name: The Spitfire and Hurricane
Addr: The Airfield, Manston Road, Ramsgate, Kent CT12 5DF
Tel: +44 (0) 1843 821940
Web: http://www.spitfire-museum.com/index.htm
Opening/Closing: 10am - 5pm May - September 10am - 4pm October - April

Name: The Royal Air Force Museum, Hendon
Addr: North London
Tel: +44 (0)20-8205 2266
Web: http://www.rafmuseum.org.uk/flashframe.cfm
Opening/Closing: Open daily from 10am to 6pm (except 1 Jan, 24-26 Dec)

Name: U.S.A.A.F. 381st Bomb Group (H) and 90 Squadron
Addr: Ridgewell, England
Tel: 01787 277310
Web: http://www.jtennet.freeserve.co.uk/
Opening/Closing: N/A

Name: Solway Aviation Museum
Addr: Crosby-on-Eden, Carlisle, CA6 4NW
Tel: +44 (0) 1228 573823
Web: http://www.btinternet.com/~lake.district/car/solavmus.htm
Opening/Closing: Open 10.30 am - 5 pm Sundays, Easter - October

Name: Southampton Hall of Aviation
Addr: Albert Road South, Southampton, Hampshire
Tel: +44 (0) 8063 5830
Web: http://www.hants.gov.uk/leisure/museums/hallavia/
Opening/Closing: Call for times

Name: Tangmere Military Aviation Museum
Addr: Tangmere , Chicheste , West Sussex , PO20 6ES
Tel: +44 (0) 1243 775 888
Web: http://www.information-britain.co.uk/attractions/tangmere/index.htm
Opening/Closing: February - November times not available.

Name: Ulster Folk and Transport Museum
Addr: Cultra, Holywood, Co.Down, BT18 0EU , Northern Ireland
Tel: +44 (0) 2890 428 428
Web: http://www.nidex.com/uftm/
Opening/Closing: Year round, call for times

Name: North East Aircraft Museum
Addr: Old Washington Road, Sunderland SR5 3HZ
Tel: (0191) 519 0662
Web: http://www.vforce.co.uk/
Opening/Closing: Daily 11am-4pm (5pm in the Summer)

Name: Wellington Aviation Museum
Addr: British School House, Broadway Road, Moreton-in-Marsh, Gloucestershire, GL56 0BG.
Tel: (01608) 650323
Web: http://www.wellingtonaviation.org/docs/index.htm
Opening/Closing: 10.00 am - 5.00 pm (12.30 - 2.00 pm Closed for Lunch) Mondays and Xmas Day Excluded

Name: The Yorkshire Air Museum
Addr: Yorkshire Air Museum, Halifax Way, Elvington, York Y041 4AU, UK.
Tel: (44) (0) 1904 608595
Web: http://www.yorksairmuseum.freeserve.co.uk/
Opening/Closing: Weekdays 11am to 3pm

Museums in the United States of America

Name: Eighth Air Force Heritage Museum
Addr: 175 Bourne Ave, Pooler, GA 31322, USA
Tel: +1 (912) 748-8888
Web: http://www.mighty8thmuseum.com/
Opening/Closing: 7 days a week from 9:00 a.m. through 6:00 p.m.

Name: 82d Airborne Division Museum
Addr: Fort Bragg at the intersection of Ardennes and Gela Streets, NC, USA
Tel: +1 (910) 432-5307
Web: http://www.fayetteville.net/museum/82ndAirborne/
Opening/Closing: TUES - SAT 10:00 am to 4:30 SUNDAY 11:30 am to 4:00 p.m. CLOSED MONDAY

Name: Air Heritage
Address: Beaver County Airport, Beaver County, PA 15010
Tel: +1 (724) 843-2820
Web: http://trfn.clpgh.org/ah/
Opening/Closing: 10:00 AM - 5:00 PM Monday - Saturday 11:00 AM - 6:00 PM Sunday

Name: Air Mobility Command Museum
Addr: 1301 Heritage Road, Dover Air Force Base, DE 19902-8001
Tel: +1 (302) 677-5938
Web: http://amcmuseum.org/popwin.html
Opening/Closing: Daily 9am-4pm

Name: Air Victory Museum
Addr: 68 Stacy Haines Rd, Medford, NJ 08055
Tel: +1 (609) 267-4488
Web: http://www.airvictorymuseum.org/
Opening/Closing: Open weekdays by appointment and Open Saturday 10:00 AM to 4:00 PM

Name: American Helicopter Museum & Education Center
Addr: 1220 American Blvd., West Chester, Pennsylvania 19380
Tel: +1 (610) 436-9600
Web: http://www.helicoptermuseum.org/
Opening/Closing: Open Wednesday - Saturday 10a.m. - 5p.m. and Sunday Noon - 5p.m.

Name: Arkansas Air Museum
Addr: Drake Field in Fayetteville, Ark
Tel: +1 (501) 521-4947
Web: http://www.arkairmuseum.org/
Opening/Closing: Mon - Fri 1:00 a.m. to 4:30 p.m.; Sat 10:00 a.m. to 4:30 p.m.; Sun 11a.m. to 4:30

Name: U.S. Army Aviation Museum
Addr: Fort Rucker, Alabama 36362-5134
Tel: 1-888-ARMYAVN
Web: http://www.aviationmuseum.org/
Opening/Closing: Monday - Friday from 9 AM - 4 PM

Name: Captain Michael King Smith Evergreen Aviation Educational Institute
Addr: 3850 Three Mile Lane McMinnville, OR 97128-9496
Tel: +1 (503) 472-9361
Web: http://www.sprucegoose.org/
Opening/Closing: open Friday through Sunday (weather permitting), from 10:00 a.m. to 5:00 p.m

Name: Castle Air Museum
Addr: Castle Air Museum is located in San Joaquin Valley, CA, adjacent to Castle Airport
Tel: +1 (209) 723-2177
Web: http://www.elite.net/castle-air/
Opening/Closing: Memorial Day thru October 1: 9:00 AM to 5:00 PM

Name: The Cavanaugh Flight Museum
Addr: 4572 Claire Chennault, Addison, TX 75001
Tel: +1 (972) 380-8800
Web: http://www.cavanaughflightmuseum.com/
Opening/Closing: Monday-Saturday from 9:00am to 5:00pm and on Sunday from 11:00am to 5:00pm

Name: Champlin Aircraft Museum
Addr: 4636 Fighter Aces Drive, Mesa, AZ 85215.
Tel: +1 (602) 830-4540
Web: http://www.arizonaguide.com/champlin
Opening/Closing: Daily 10 to 5

Name: Dare County Regional Airport Museum
Addr: 410 Airport Road, Manteo, North Carolina 27954
Tel: +1 (252) 473.2600
Web: http://www.flymqi.com/museum/
Opening/Closing: 8am till 7pm 7 days a week

Name: EAA Aviation Center
Address: 3000 Poberezny Road, Oshkosh, WI 54903-30
Tel: +1 (920) 426-4800
Web: http://museum.eaa.org/
Opening/Closing: Monday-Saturday 8:30 a.m. to 5 p.m. Sunday 11 a.m. to 5 p.m.

Name: The Flying Tigers Warbird Restoration Museum
Addr: 231 North Hoagland Boulevard, Kissimmee, FL 34741
Tel: +1 (407) 933-1942
Web: http://www.warbirdmuseum.com/
Opening/Closing: Monday - Saturday 9:00 am to 5:00 pm Sunday 9:00 am to 5:00 pm

Name: The Freedom Museum
Addr: 10400 Terminal Road, Manassas, Virginia 20110
Tel: +1 (703) 393-0660
Web: http://www.freedommuseum.org/
Opening/Closing: Monday - Saturday 9 am to 5 pm. Sundays 12 - 4 pm.

Name: The Freedom Museum
Addr: 8008 Cedar Springs Rd. Mockingbird Ln. @ Cedar Springs Rd.
Tel: +1 (214) 350-1651
Web: http://flightmuseum.com/
Opening/Closing: Mon.-Sat. 10am to 5pm, Sunday 1pm to 5pm

Name: The Glenn L. Martin Aviation Museum
Addr: Martin State Airport, Middle River, Maryland, near Baltimore.
Tel: +1 (410) 682-6122
Web: http://www.martinstateairport.com/museum
Opening/Closing: Monday through Friday 10:00 AM to 2:00 PM and Saturday 1:00 PM to 5:00 PM

Name: Hanger 10 Flying Museum
Addr: 1945 Matt Wright Lane, Denton Municipal Airport, Denton, Tx 76207
Tel: +1 (940) 565-1945
Web: http://www.hangar10.org/
Opening/Closing: N/A

Name: Hill Aerospace Museum
Addr: 75th ABW/MU, 7961 Wardleigh Rd, Hill AFB, UT 84056-5842
Tel: +1 (801) 777-6868
Web: http://www.hill.af.mil/museum/
Opening/Closing: Seven days a week from 9:00 AM til 4:30 PM.

Name: Hiller Aviation Museum
Addr: 601 Skyway Road, San Carlos, CA. 94070
Tel: +1 (650) 654-0200
Web: http://www.hiller.org
Opening/Closing: 10am-5pm, 7 days a week

Name: International Sport Aviation Museum
Addr: P.O. Box 6795, Lakeland, FL 33807-6795
Tel: +1 (863) 644-0741
Web: http://www.airmuseum.org/
Opening/Closing: Monday - Friday: 9 a.m. - 5 p.m. Saturday: 10 a.m. - 4 p.m. Sunday: Noon - 4 p.m.

Name: Kansas Aviation
Addr: 3350 George Washington Boulevard, Wichita, Kansas 67210
Tel: +1 (316) 683-9242
Web: http://www.saranap.com/kam.html
Opening/Closing: 9 A.M. to 4 P.M. Tuesdays-Fridays; 1 P.M. to 5 P.M. Saturdays

Name: Hall of Space Museum
Addr: 1100 N. Plum Hutchinson, Kansas 67501
Tel: +1 (316) 662-2305
Web: http://www.cosmo.org/
Opening/Closing: Friday and Saturday: 4 pm Sunday: 2 pm

Name: MAPS Air Museum
Addr: 5359 Massillon Road, North Canton, Ohio 44720
Tel: +1 (330) 896-6332
Web: http://www.mapsairmuseum.org/
Opening/Closing: Monday: 9:00am to 4:00pm Wednesday: 6:00pm to 9:00pm Saturday: 8:00am to 4:00pm

Name: Mid Atlantic Air Museum
Addr: 11 Museum Drive, Reading, Pennsylvania 19605
Tel: +1 (610) 372-7333
Web: http://www.maam.org/
Opening/Closing: Open daily 9:30 AM - 4:00 PM

Name: Mid America Air Museum
Addr: 2000 West 2nd, Liberal, KS 67901
Tel: +1 (316) 624-5263
Web: http://www.airmuseum.net/
Opening/Closing: Mon.-Sat. 10-5pm, Sun. 1-5pm

Name: Minnesota Air Guard Museum
Addr: ANG Base - Mpls-St.Paul International Airport, St. Paul, MN 55111-0598
Tel: +1 (612) 713-2523
Web: http://www.mnangmuseum.org/index.html
Opening/Closing: Saturdays and Sundays 11:00 am to 4:00 p.m.

Name: The Museum of Aviation
Addr: Warner Robins, GA.
Tel: +1 (478) 926-6870
Web: http://www.museumofaviation.org/
Opening/Closing: Open seven days a week from 9am to 5pm

Name: Seattle's Museum of Flight
Addr: METRO bus #174 from downtown Seattle or Sea-Tac Airport
Tel: +1 (206) 764-5720
Web: http://www.museumofflight.org/
Opening/Closing: Open daily: 10am to 5pm Thursdays: 10am to 9pm

Name: National Air and Space Museum
Addr: 7th and Independence Ave, SW, Washington, DC 20560
Tel: +1 (202) 357-2700
Web: http://www.nasm.edu/
Opening/Closing: Daily 9:30 a.m. - 5:30 p.m

Name: National Museum of Naval Aviation
Addr: Naval Air Station Pensacola, Florida
Tel: +1 (850) 452-3604
Web: http://www.naval-air.org/Museum_Home_Frame.htm
Opening/Closing: Open 9 AM to 5 PM Daily

Name: National Soaring Museum
Addr: 51 Soaring Hill Drive, Elmira, NY 14903-9204
Tel: +1 (607) 734-3128
Web: http://www.soaringmuseum.org/
Opening/Closing: Open 7 Days; 10 am to 5 pm

Name: National Warplane Museum
Addr: 17 Aviation Drive, Horseheads, NY 14845
Tel: +1 (607) 739-8200
Web: http://www.warplane.org/
Opening/Closing: Monday - Saturday: 9:00a.m.-5:00p.m. Sunday: 11:00a.m.-5:00p.m.

Name: New England Air Museum
Addr: Bradley International Airport, Windsor Locks, CT 06096
Tel: +1 (860) 623-3305
Web: http://www.neam.org/
Opening/Closing: Open 7 days a week, year round, 10am - 5pm

Name: Palm Springs Air Museum
Addr: 745 North Gene Autry Trail, Palm Springs, California 92262
Tel: +1 (760) 778-6262
Web: http://www.air-museum.org/
Opening/Closing: Open seven days a week, year round, from 10:00 am to 5:00 pm.

Name: Pima Air and Space Museum
Addr: 6000 East Valencia Road , Tucson, AZ 85706
Tel: +1 (520) 574-0462
Web: http://www.pimaair.org
Opening/Closing: N/A:

Name: Piper Aviation Museum
Addr: Lock Haven, Clinton County, Pennsylvania, U.S.A.
Tel: +1 (570)-748-8283
Web: http://www.kcnet.org/~piper/
Opening/Closing: Mon - Frid 9:00 AM to 4:30 PM Sat 10:00 AM to 4:00 PM Sun 12 Noon to 4:00 PM

Name: San Diego Aerospace Museum
Addr: 2001 Pan American Plaza, Balboa Park, San Diego, CA 92101
Tel: +1 (619) 234-8291
Web: http://www.aerospacemuseum.org/
Opening/Closing: 10 a.m. to 4:30 p.m. with last admission taken at 4:00 p.m.

Name: The Virginia Air & Space Center
Addr: 600 Settlers Landing Road, Hampton, VA, 23669-4033
Tel: +1 (757) 727-0900
Web: http://www.vasc.org/
Opening/Closing: Winter, Mon-Sun 10:00 a.m. to 5:00 p.m. Summer,. Mon-Wed 10:00 a.m. to 5:00 p.m.

Name: The Virginia Aviation Museum
Addr: Richmond International Airport at 5701 Huntsman Road in Sandston, Virginia.
Tel: +1 (804) 236-3622
Web: http://www.smv.org/
Opening/Closing: Open daily, except Thanksgiving and December 25, from 9:30 a.m. to 5 p.m.

Name: WACO Air Museum
Addr: Troy, Ohio
Tel: +1 (937) 335-WACO
Web: http://www.wacoairmuseum.org/
Opening/Closing: May - Oct; Sat. & Sun 1pm - 5 p.m

Name: The War Eagles Air Museum
Addr: 8012 Airport Rd, Santa Teresa, NM 88008
Tel: +1 (505) 589-2000
Web: http://www.war-eagles-air-museum.com/
Opening/Closing: N/A

Name: Western Museum of Flight
Addr: 12016 Prairie Avenue, Hawthorne, California 90250
Tel: +1 (310) 332-6228
Web: http://www.wmof.com/
Opening/Closing: Tuesday through Saturday, from 10:00 am to 3:00 pm

Name: The Wings of Freedom
Addr: Huntington Municipal Airport, 1365 Warren Road, Huntington, Indiana 46750
Tel: 219-356-1945
Web: http://hometown.aol.com/scatvii/page/index.htm
Opening/Closing: Sat 10-4, Sun 7-4

Name: The Wings of Freedom
Addr: Huntington Municipal Airport, 1365 Warren Road, Huntington, Indiana 46750
Tel: 219-356-1945
Web: http://hometown.aol.com/scatvii/page/index.htm
Opening/Closing: Sat 10-4, Sun 7-4

Name: The Wings Of History Air Museum
Addr: 12777 Murphy Avenue, San Martin, California USA
Tel: 408-683-2290
Web: http://www.wingsofhistory.org/museum.htm
Opening/Closing: Saturday and Sunday 11am - 4pm

Name: The Wings Over The Rockies Air and Space Museum
Addr: Hangar Number 1, 7711 East Academy Boulevard, Denver, CO 80220-6929 USA
Tel: 303-360-5360
Web: http://www.dimensional.com/~worm/
Opening/Closing: Mon-Sat 10AM to 5PM, Sun 12 Noon to 5PM

Aircraft Manufacturers

Aircraft Manufacturers

A

21st Century Airships Inc	http://21stcenturyairships.com
Aasi - Advanced Aerodynamics and Structures Inc	http://www.aasiaircraft.com
ABC - American Blimp Corporation	http://www.americanblimp.com
Ad Aerospace Ltd	http://www.ad-aero.co.uk/
Ada - Aeronautical Development Agency	http://www.ada.gov.in/
Aero Vodochody As	http://www.aero.cz
Aerocomp	http://www.aerocompinc.com
AEROS - Worldwide Aeros Corporation	http://www.aeros-airships.com
Aerostar - Aerostar Aircraft Corporation	http://www.aerostaraircraft.com
AIDC Aerospace Industrial Dev'ment Corporation	http://www.aidc.com.tw/
Air Tractor Inc	http://www.airtractor.com
Airbus Industrie	http://www.airbus.com
Aircraft Designs Inc	http://www.aircraftdesigns.com
Airship Technologies Europe Ltd	http://www.airship.com
Alberta Aerospace Corporation	http://www.aerospace.ab.ca/
Alenia Aerospazio (a Finmeccanica Company)	http://www.alespazio.it
AMD - Aircraft Manufacturing & Development	http://www.newplane.com
American Champion Aircraft	http://www.amerchampionaircraft.com
American Homebuilts' Inc	http://www.americanhomebuilts.com
American Sportscopter Inc	http://www.ultrasport.rotor.com
American Utilicraft Corporation	http://www.utilicraft.com
Aquila Technische Entwicklungen Gmbh	http://www.aquila-aero.com
Atr - Avions De Transport Regional	http://www.ataircraft.com
Aviat Aircraft Inc	http://www.aviataircraft.com
Avid Aircraft Inc	http://www.avidair.com/
Avro	http://www.bae.co.uk/
Ayres Corporation	http://www.ayrescorp.com

B

Bae Advanced Aircraft Studies	http://www.bae.co.uk/
Bae Systems (operations) Ltd	http://www.bae.co.uk/
Bae Systems Airbus	http://www.bae.co.uk/
Barr Aircraft	http://www.barraircraft.com/
Beijing University Of Aeronautics and Astronautics	http://www.buaa.edu.cn/
Bell Helicopter Textron Canada	http://www.bellhelicopter.textron.com
Bell Helicopter Textron Inc	http://www.bellhelicopter.textron.com
Bell/agusta Aerospace Company	http://www.bellagusta.com/
Bharat Heavy Electricals Ltd	http://www.bhel.com/
Boeing Business Jets	http://www.boeing.com
Boeing Company and Sikorsky Aircraft	http://www.boeing.com
Boeing/bae	http://www.boeing.com
Bombardier Aerospace	http://www.learjet.com
Bombardier Aerospace Canadair	http://www.learjet.com
Bombardier Aerospace De Havilland	http://www.learjet.com
Bombardier Aerospace Learjet	http://www.learjet.com
Brantly International Inc (helicopter Division)	http://www.brantly.com/
Britten-norman Ltd	http://www.britten-norman.com/

C

Cameron Balloons Ltd	http://www.cameronballoons.co.uk
Cap Aviation	http://www.capaviation.com
Cargolifter Ag	http://www.cargolifter.com/
Carlson Aircraft - Carlson	http://www.sky-tek.com/
CASA - Construcciones Aeronauticas Sa	http://www.casa.es/
Century Aerospace Corporation	http://www.centuryaero.com/
Cessna Aircraft Company	http://www.cessna.textron.com
Cfm Aircraft	http://www.cfm-aircraft.co.uk/
Cirrus Design Corporation	http://www.cirrusdesign.com/
Commander Aircraft Company	http://www.commanderair.com/
Culp's Specialties	http://www.culpsspecialties.com
Czech Aircraft Works S.r.o	http://www.zenithair.com

D

Daimlerchrysler Aerospace Airbus/tupolev	http://www.dasa.com/
Dasa	http://www.dasa.com/
Dassault Aviation	http://www.dassault-aviation.com
Delta System-air As	http://www.dsa.cz
Denel Aviation (division Of Denel (pty) Ltd)	http://www.denel.co.za/
Derringer Aircraft Company Llc	http://www.derringeraircraft.com/
Diamond Aircraft Industries Gmbh	http://www.diamondair.com
Dornier Luftfahrt Gmbh	http://www.fairchildaerospace.com/
Dreamwings Llc	http://www.dreamwings.com/

E

Eagle Aircraft Pty Ltd	http://www.eagleair.com.au/
Eclipse Aviation Corporation	http://www.eclipseaviation.com
El Gavilán Sa	http://www.elgavilan.com/
Empresa Nacional De Aeronáutica De Chile	http://www.enaer.cl/
Eurocopter (eurocopter France)	http://www.eurocopter.com
Eurocopter Canada Ltd	http://www.eurocopter.com
Eurocopter/catic/st Aero	http://www.eurocopter.com
Eurocopter/kawasaki	http://www.eurocopter.com
Euro-enaer Holding Bv	http://www.eurocopter.com
Eurofighter Typhoon	http://www.eurofighter-typhoon.com
Europa Aviation Ltd	http://www.europa-aircraft.com
Express Aircraft Company	http://www.express-aircraft.com
Extra-flugzeugbau Gmbh	http://www.extraflugzeugbau.com

F

Fairchild Aerospace Corporation	http://www.fairchildaerospace.com
Fairchild Dornier	http://www.fairchildaerospace.com/
Farnborough-aircraft.com	http://www.farnborough-aircraft.com
Fisher Flying Products	http://www.fisherflying.com/
Found Aircraft Canada Inc	http://www.foundair.com/

G

Gkn Westland Helicopters Limited	http://www.gkn-whl.co.uk/
Goair Products	http://www.bankstownairport.com.au
Great Plains Aircraft Supply Co Inc	http://www.greatplainsas.com/
Griffon Aerospace	http://www.griffon-aerospace.com/
Groen Brothers Aviation Inc	http://www.groenbros.com/
Gulfstream Aerospace Corporation	http://www.gulfstreamaircraft.com
Gsi - Global Skyship Industries Inc	http://www.globalskyships.com

H

HAL - Hindustan Aeronautics Limited	http://www.hal-india.com

I

Iai - Israel Aircraft Industries Ltd	http://www.iai.co.il/
Ibis Aerospace Ltd	http://www.ibisaerospace.com/
Iniziative Industriali Italiane Spa	http://www.xs4all.nl/~ronair/arrow.html
Instytut Lotnictwa (aviation Institute)	http://www.ilot.edu.pl/
Israel Aircraft Industries Ltd	http://www.iai.co.il/

J

Jabiru Aircraft Pty Ltd	http://www.jabiru.net.au
Jammaero/starcorp	http://www.jammaero.com
Jim Kimball Enterprises Inc	http://www.jimkimballenterprises.com/

K

Kaman Aerospace Corporation	http://www.kamanaero.com/index.htm
Kestrel Aircraft Company	http://www.kestrelaircraft.com/

L

Lancair Group Inc	http://www.lancair.com/
Lindstrand Balloons Ltd	http://www.lindstrand.co.uk/
Lockheed Martin Aircraft Argentina Sa	http://www.lockheedmartin.com
Lockheed Martin Corporation	http://www.lockheedmartin.com
Lopresti Inc	http://www.LoPrestiFury.com/
Luscombe Aircraft Corporation	http://www.luscombeaircraft.com

M

Magni Gyro	http://www.magnigyro.com
Masquito Aircraft N.v	http://www.masquito.be
Maule Air Inc	http://www.mauleairinc.com
Maverick Air Inc	http://www.maverickhelicopter.com
Md Helicopters	http://www.md900.com
Micco Aircraft Company	http://www.miccoair.com
Mitsubishi Jukogyo Kabushiki	http://www.mhi.co.jp
Mooney Aircraft Corporation	http://www.mooney.com
Murphy Aircraft Manufacturing Ltd	http://www.murphyair.com
Mylius Flugzeugwerk Gmbh & Co Kg	http://www.mylius-flugzeugwerke.de

N

Nal - National Aerospace Laboratories
National Aerospace Laboratory
National Space Development Agency Of Japan
New Kolb Aircraft Company Inc
Northrop Grumman Corporation
Nusantara Aircraft Industries Ltd

http://www.cmmacs.ernet.in
http://spaceboy.nasda.go.jp
http://www.nasda.go.jp
http://www.tnkolbaircraft.com
http://www.northgrum.com
http://www.iptn.co.id

P

Pakistan Aeronautical Complex
Papa 51 Ltd
Paxman's Northern Lite Aerocraft
Pilatus Flugzeugwerke Ag
Pulsar - Pulsar Aircraft Corporation

http://www.pac.org.pk
http://www.thundermustang.com
http://www.lis.ab.ca/paxman
http://www.pilatus-aircraft.com
http://www.pulsaraircraft.com

Q

Quikkit - Rainbow Flyers Inc - Quikkit Division

http://www.glassgoose.com

R

Raf - Rotary Air Force Inc
Rans Inc
Raytheon - Raytheon Aircraft Company
Raytheon Aircraft Company
Reims Aviation Sa
Rigid Airship Design Nv - Rigidair
Robinson Helicopter Company
Rotary Air Force Inc
Rotorway International

http://www.raf2000.com
http://www.rans.com
http://www.raytheon.com
http://www.raytheon.com
http://www.reims-aviation.fr
http://www.rigidair.com
http://www.robinsonheli.com
http://www.raf2000.com
http://www.rotorway.com

S

Saab Ab
Safire Aircraft Company
Samara State Aerospace University
Scaled Composites Inc
Schweizer Aircraft Corporation
Seastar Aircraft Inc
Sequoia Aircraft Corporation
Sg Aviation
Sikorsky Aircraft
Sino Swearingen Aircraft Company
Skystar Aircraft Corporation
Slingsby Aviation Limited
Sme Aviation Sdn Bhd
Socata Group Aerospatiale
Stemme Gmbh & Co Kg
Stoddard-hamilton Aircraft Inc
Super-chipmunk

http://www.saab.se
http://www.safireaircraft.com
http://www.ssau.ru/engl/
http://www.scaled.com
http://www.schweizer-aircraft.com
http://www.seastarplane.com
http://www.SeqAir.com
http://www.storm-sg.it
http://www.sikorsky.com
http://www.sj30jet.com
http://www.skystar.com
http://www.slingsby.co.uk
http://www.smeav.com.my/
http://www.socata.com
http://www.motorglider.com
http://www.stoddard-hamilton.com
http://www.super-chipmunk.com

T

Taneja Aerospace and Aviation Ltd	http://www.tanejaaerospace.com
Technoflug Leichtflugzeugbau Gmbh	http://www.technoflug.de
Tecnam - Costruzioni Aeronautiche Tecnam Srl	http://www.tecnam.com
The Boeing Company	http://www.boeing.com
The Enstrom Helicopter	http://www.enstromhelicopter.com
The Hamilton Airship Company	http://www.hamilton.co.za/
The New Piper Aircraft Inc	http://www.newpiper.com
Turkish Aerospace Industries Inc	http://www.tai.com.tr/

V

Van's Aircraft Inc	http://www.vansaircraft.com
Visionaire Corporation	http://www.visionaire.com
Vstol Aircraft Corporation	http://www.vstolaircraft.com
Vulcanair Spa	http://www.vulcanair.com

W

Waco Classic Aircraft Corporation	http://www.wacoclassic.com

Z

Zenith Aircraft Company	http://www.zenithair.com
Zeppelin Luftschifftechnik Gmbh	http://www.zeppelin-nt.com
Zivko Aeronautics Inc	http://www.zivko.com

Flight Simulation
Web Sites

A selection of web sites in flight simulation. As this book is limited in size obviously we cannot include all that are found on the Internet but as you explore you will discover many, many more.

Add-ons / Downloads

Absolute Flight Sim	http://home.onet.co.uk/~rdavis/fs/
AeroNet	http://fly-aeronet.com
AFDA - Professional adventures for MSFS	http://www.afda.de
Airline Manager	http://www.airline-manager.atfreeweb.com
AirNetherlands Virtual Airline	http://www.fly.to/airnetherlands/
Alaska By Don	http://www.avsim.com/alaska
Andy's Flight Simulator Page	http://www.crosswinds.net/~andysflightsim
Austrian Aircraft Factory	http://www.flightsimmers.net/aaf
Chris' Airbus Site	http://www.cwbalmer.co.uk
Clearance Unlimited	http://www.clearanceunlimited.com
Col's Melbourne CrewRoom	http://www.geocities.com/colinlock.geo/
Ecuadorian add-ons for FS98/FS2000	http://www.geocities.com/ecuadorfs98
Eric C. Johnson's Simply Planes Page	http://flightsimmers.net/ecjohnson
Flight Sim Software	http://www.flightsimsoftware.co.uk
Flight Simulator Nordic	http://www.fsnordic.net
Flight Simulator utils uk	http://www.bentley55.freeserve.co.uk
FlightBase2000	http://www.flightbase2000.com
Flightlabs.com	http://www.flightlabs.com
FlightSim.Com	http://www.flightsim.com/
FlightSimUK	http://website.lineone.net/~flightsimuk
FlightSim-uk.com	http://www.flightsim-uk.com
Flightsimulator Users Group	http://www.fsgg.nl
Flugplätze für Deutschland	http://flightsimmers.net/testvt
FS 98 Addons	http://FlightSim98Addons.cjb.net
FS2000 Best of the Best	http://www.users.totalise.co.uk/~harris_rob/
FSfreeware	http://www.fsfreeware.com
FSUK-Downloads	http://www.fsuk-downloads.co.uk
FSZwever	http://www.flightsimmers.net/fszwever/
H_Squad Ready Room	http://www.navyair.com
Helix FS	http://helix_fs.tripod.com
Horizons Virtuels	http://www.horizons-virtuels.com/
Jetstream 41	http://www.jetstream.41.btinternet.co.uk
MaxTac's Aircrafts for MSFS	http://www.maxtac.it
Michael's Wisconsin Flight Simulator Page	http://www.geocities.com/CapeCanaveral/Hangar/4925/
Mick's Flight Sim Site	http://members.nbci.com/mickwilliams/
Mirage Aircraft for Flight Simulator	http://www.flightsimmers.net/airbase/mirage
Motion Flight Simulator for FS 2000	http://www.flightsimulator.ch
NordicNav	http://w1.321.telia.com/~u32104602/nordicnav/
Norwood Aerospace Industries FS	http://www.flightsimmers.net/airbase/norwoodae
Paradise Helipad	http://www.paradisehelipad.crosswinds.net
Pat's Flight Simulator Site	http://digilander.iol.it/Migliola/P_Migliola.htm
Paul's Freeware Military Panels	http://www.flightsimmers.net/airbase/paulsmp
Perfect Flight 2000 project	http://www.fs2000.org

PJD Software	http://www.pjdsoftware.co.uk
Planesimulation	http://www.planesimulation.com/
Premier Aircraft Design	http://www.flightsimnetwork.com/premaircraft/home.htm
Professional Flight Displays,Inc.	http://www.pfdteam.com/
Schiratti.Com	http://www.schiratti.com
Shigeru's Aircraft Models	http://www.flightsimnetwork.com/shigeru/
sim stuff	http://www.fsaircarft.com
Sim Utilities Gold	http://flightsimmers.net/aindex/index.html
SimAir 2000	http://www.simair2000.com
SimFlight Argentina	http://www.simflight.com.ar
SimFlight México	http://www.simflight.com/mexico
Simvol	http://www.simvol.org
six33 Squadron (CFS)	http://www.six33.co.uk
SurClaro.com	http://www.surclaro.com/
Swedflight - Sceneries for Sweden	http://www.flightsim.no/swedflight
The DC-8 Project Team	http://www.flightsimnetwork.com/dc8_team/home.htm
The Freeware Works	http://www.freewareworks.com
The Norwegian FlightSim Hangar	http://home.online.no/~alfjoha/
Tthe Pakistan flightsim page	http://www.geocities.com/pakistanflightsim
The Pre-Flight Checklist Website	http://www.geocities.com/flight_simmer/
The Scenery Hall of Fame	http://www.erols.com/tdg
Virtual FAA	http://www.vfaa.org
Virtual Flights	http://intercity.it/moneta
World of FS	http://www.worldoffs.com

Books / Magazines

Flight Simulator World Magazine	http://www.flightsimulatorworld.com/
PC Aviator / Computer Pilot	http://pcaviator.com.au/
PC Pilot	http://www.pcpilot.net
TopSkills	http://www.topskills.com/flitsim.htm
TecPilot Publishing	http://www.tecpilot.com

Combat Simulation (Microsoft Related)

CFS2 Online	http://cfs2.dogfighter.com
CFSUK	http://www.oasts.btinternet.co.uk/cfsmain.htm
Combat Flight Center	http://www.combatfs.com
MiGMan's Flight Sim Museum	http://www.migman.com
Sim Arena	http://www.sim-arena.com
Sparkys Hangar	http://sparkys-hanger.is-crazy.com
The Aerodrome	http://www.rusty100.freeserve.co.uk/the%20aerdrome

Developers

Aircraft3-view	http://www.flightsimnetwork.com/aircraft3-view
AlliedFsGroup	http://www.alliedfsgroup.com
Euroflight	http://www.avsim.com/euroflight
FGA Network	http://www.fganetwork.com
FS Design Studio Developer Pages	http://www.fsdesignstudio.com
FS DREAM FACTORY	http://m1.313.telia.com/~u31309548/
GB Airports	http://www.gbairports.co.uk
JORGE SANTORO'S HOME PAGE	http://users.task.com.br/santoro/
MMD	http://www.geocities.com/marfelde
Stefan2000	http://www.stefan2000.fly.to/
The Frederic Nadot's site	http://perso.club-internet.fr/fnadot
Touchdown Software	http://www.touchdown-software.com
UK2000 Scenery	http://www.uk2000scenery.co.uk

FS Shopping

Flight Sim Central	http://www.flightsimcentral.com
FS Share Center	http://www.fssharecenter.com
RC Simulations	http://www.rcsimulations.com
SimMarket	http://www.simmarket.com
Virtual Reality Ltd	http://www.virtualrealityuk.com

Links / Search

Mike's Airport	http://www.angelfire.com/ca5/flightsim81
Pegleg Productions	http://www.peglegs.net
The Fly! Links Site	http://www.avsim.com/flylinks
The Greatest Site of Links Ever!	http://www.BestLinkSite.com
Visual Approach	http://members.home.net/aviation/

Multiplayer / ATC

Aircom	http://www.aircom.org
DK-VACC	http://www.fly.opasia.dk/
Halcones Virtuales	http://www.halconesvirtuales.com
Radar Contact	http://www.flightsimmers.net/radarcontact
The IVAO	http://www.ivao.org/
The KA7 War Machine	http://home.golden.net/`jackstraw
The Top Fighters Group	http://www.geocities.com/the_top_fighters_group/frames_central.html
VATSIM - (MAIN SITE)	http://www.vatsim.net
VATSIM-EUR	http://www.vatsim-eur.org
VATSIM-UK	http://www.vatsim-uk.net
VATSIM-USA	http://www.vatusa-org

News / Information

A320 Project	http://www.avsim.com/hangar/flight/a320project/main.html
Avsim Online	http://www.avsim.com
FlightSim.com	http://www.flightsim.com
FlightSims.co.uk	http://www.flightsims.co.uk
FS2000 with Capt. Koert	http://koerte.com/fs2000
GA-Sim	http://www.ga-sim.com
Hangar 21	http://www.emportugal.com/in/hangar21
John W Nobes Homepage	http://www.jw.nobes.btinternet.co.uk
SimFlight Network	http://www.simflight.com
Takeoffpro	http://www.takeoffpro.de
TecPilot	http://www.tecpilot.com
The Fly! Showcase	http://www.saniku.ac.jp/images/fly/flyshowcase.htm

Pictures

AirWeb - FS Images from around the World	http://www.teleport.com/~ksamco/AirWeb/index.html
FlightSim.com's Virtual Fly-In	http://vfi-photos1.bigsitecity.com
Fly to Fsflight	http://fly.to/fsflight
The PMDG Aircraft Photo Gallery	http://www.saniku.ac.jp/images/pmdg/gallery.htm

Publishers

Abacus	http://www.abacuspub.com
Aerosoft	http://www.aerosoft.com
Alpha Simulations	http://www.alphasim.co.uk/
Apollo Software	http://www.apollosoftware.com/
CH products	http://www.chproducts.com
Flight 1	http://www.flight1.com
Just Flight	http://www.justflight.com
Lago Srl	http://www.lagoonline.com
Microsoft (main site)	http://www.microsoft.com
SimMarket	http://www.simmarket.com
TecPilot Publishing	http://www.tecpilot.com
UK2000 Scenery Project	http://www.uk2000scenery.com/puplic/index.htm
Wilco Publishing	http://www.wilcopub.com

"Real Aviation"

Costa Rican Aviation	http://fly.to/sjo
Culpeper Aero Squadron	http://www.gtechno.com/cas
Danish Acars Homepage	http://www.acars.subnet.dk
Euroairport Basel-Mulhouse (LFSB)	http://koerte.com/lfsb
The DC3 Hangar	http://www.douglasdc3.com

Training / Guidance

747 Simulator	http://www.hyway.com.au/747/747.html
Al's Flightsim Site	http://members.aol.com/alhb/index.html
Flight-Sim.net	http://www.flight-sim.net
Freeflight Design Studio	http://www.freeflightdesign.com
FSTA	http://fsta.iwarp.com
HoverSafe Academy	http://home.iprimus.com.au/pwinwoo
IFR Plans	http://www.ifrplans.com
Navigation Tutorial for Flight Simulation Fans	http://www.navflightsim.com
Rick Lee's FS Page	http://www.rickleephoto.com/rlfs.htm
Virtual FAA	http://www.vfaa.org

Virtual Airlines

60th HSU (Helicopter Support Unit)	http://www.geocities.com/60thvaw/60thhsu/index.html
Aerolineas Argentinas Virtual	http://www.aerolineasvirtual.com.ar
African International Airlines	http://www.evalunet.co.za/evalunet/aia/index.htm
Air Canadian	http://www.aircdn.com
Air Kansas	http://www.llarson.com/airks
Air Pacifica Virtual Airlines	http://www.peggware.com/airpacifica/
AirFloirda Virtual Airlines	http://www.flyairflorida.com
AirTran Virtual Airways	http://www.airtranva.com
AirWeGo Virtual Airline Group	http://www.airwego.org.uk/
AirWest Virtual Airlines	http://www.flyairwest.com
American Star Airlines	http://www.americanstarair.com
AmEuro Virtual Airlines	http://members.home.net/ameuro
ATCWorldStar Alliance	http://www.atcworldstar.co.uk
Atoll Airways	http://www.polarairways.com
Australian Colonial Airways (VA)	http://www.acavirtualairways.org
Benair Virtual Airlines	http://www.8op.com/benair
Blue Yonder Aviation	http://www.blueyonder.de
Bluegrass Airlines GW2000	http://www.dslky.net/users/popawjoe
Braniff International Virtual Airways	http://members.tripod.com/BIVA_2/BIVA.html
British Airways Virtual	http://www.bavirtual.co.uk
British Atlantic Virtual Airlines	http://www.geocities.com/britishatlantic
Caesar Air 2000	http://www.caesar-air.com
Caesar Air Worlwide VA	http://www.caesar-air.com
Canadian Virtual Airlines	http://www.canadianva.com
Cardinal Airlines	http://www.cardam.com
Cayuga Airways	http://www.geocities.com/cayuga_airways
Celtic World Airways	http://www.geocities.com/willy515_99
Coastline Air	http://www.jschein.com/coastal.html
Continental-Northwest	http://www.cnva.cjb.net
CP Air VA	http://www.flightsimnetwork.com/cpairva/

Croatia Airlines Virtual	http://ctnvirt.cjb.net
Delaware Airlines	http://www.delawareairlines.com
Desert Air	http://www.smibdaddy.com
East Coast Air	http://kmc421.tripod.com/eastcoastair2
Europa Airways	http://www.europa-airways.de
EuropAir Virtual Airline	http://www.europairva.net
Eurostar	http://www.esava.net
Fox River Helicopters	http://www.flightsimmers.net/helipad/foxriverva/index.htm
Gatwick Virtual Air Taxi	http://www.gvat.btinternet.co.uk
German Airlines	http://www.german-airlines.com
Just-Flights Airlines	http://www.jfa.org.uk
Leeds Lines - THE Virtual Airline	http://www.leedslines.co.uk
Lufthansa Virtual airlines	http://www.lufthansavirtual.homestead.com
Maveric Virtual Airlines	http://www.mavericva.co.uk
Meridian Airlines	http://www.avsim.com/meridian
Mil-Air	http://www.members.tripod.com/~imalm/default.htm
Newfoundland Air	http://www.flightsimnetwork.com/nfair/
Noble Air	http://www.nobleair.com
Northern Star Airlines	http://www.northernstarair.com
Pacific Breeze Virtual Airlines	http://www.pacbreezeva.com
Pacific West Airways	http://www.pacificwestairways.com
Polar Airways	http://www.polarairways.com
Royal Flying Doctor Service Virtual Airline	http://rfds.virtual.tripod.com
Scottish Airways	http://www.flightsim-uk.com/sava
SKY Israel Airlines	http://www.skyisrael.cjb.net
SRAS (Snake River Air Service)	http://home.twcny.rr.com/slchub/sra.htm
SUR Air System	http://www.planetaviation.com/virtualairlines/airlines/sur
Swissair Virtual Group	http://www.ssva.cjb.net
The Battle of the Airlines Trophy	http://www.planetaviation.com/virtualairlines/battle/
The Defenders of Yorktown	http://www.johnbmayes.com
The Vitrual United States Air Forces	http://www.vusafs.com
TransVirtual Airlines	http://www.transvirtualairlines.com
Virgin International Airways	http://www.virginva.cjb.net
Virtu Alitalia	http://www.virtualitalia.it
Virtual Cardinal Airlines	http://www.cardam.com
Virtual Malaysia Airlines	http://uk.geocities.com/virtualmas/index.htm
Virtual United States Air Forces	http://www.geocities.com/c5keeper

Glossary

A

A	Autotuned NAVAID
A	At or Above (constrained altitude)
AA	American
AAATS	Australian Advanced Air Traffic Services
AAC	Aeronautical Administration Communication
AAS	Advanced Automation System
AATT	Advanced Aviation Transportation Technology
ABM	Abeam
A/C	Aircraft
AC	Air Canada
ACARS	Aircraft Communication Addressing and Reporting System
ACARS	ARINC communications & Address Reporting System
ACARS	MU ACARS Management Unit
ACAS	Airborne Collision and Avoidance System
ACF	Area Control Facility
ACFS	Advanced Concepts Flight Simulator
ACK	Acknowledge
ACMS	Aircraft Condition Monitoring System
ACT	Active
ADC	Air Data Computer
ADF	Automatic Direction Finder
ADI	Attitude Director indicator
ADLP	Aircraft Data Link Processor
ADMA	Aviation Distributors and Manufacturers Association
ADS	Automatic Dependent Surveillance
AECB	Atomic Energy Control Board
AERA	Automated Enroute ATC
AFCS	Automatic Flight Control System
AFDS	Autopilot Flight Director System (also A/P F/D)
AFS	Automatic Flight System
AGATE	Advanced General Aviation Transport Experiments
AGL	Above Ground Level
AHRS	Altitude Heading Reference System
AIRS	Advanced Infrared Sounder
A/I	Anti-ice
AI	Artificial Intelligence
AL	Allegheny
ALPA	Air Line Pilots Association
ALT	Altitude
ALT	Alternate
ALTN	Alternate
ALT	HOLD Altitude Hold Mode
AM	Amplitude Modulation
AM	Aero Mexico
AMSS	Aeronautical Mobile Satellite Service
ANA	All Nippon Airways
AOA	Angle-of-Attack
AOC	Aeronautical Operation Control
AOCS	Attitude and Orbit Control System
AOM	Aircraft Operating Manual
AOPA	Aircraft Owners and Pilots Association
A/P	Autopilot
APA	Allied Pilots Association
APC	Aeronautical Passenger Communication
APMS	Automated Performance Measurement System
APPR	Approach/Approach Mode
APR	April
APRT	Airport
APU	Auxiliary Power Unit
AQP	Advanced Qualification Program
ARAC	Aviation Rulemaking Advisory Committee
ARINC	Aeronautical Radio Incorporated
ARPA	Advanced Research Projects Agency
ARR	Arrival
ARTCC	Air Route Traffic Control Center
ARTS	Automated Radar Terminal System
ASCII	US Code for Interface and Interchange
ASI	Air Speed Indicator
ASR	Airport Surveillance Radar
ASRS	Aviation Safety Reporting System
AT	At (an altitude)
A/T	Autothrottle
ATA	Air Transport Association
ATA	Actual Time of Arrival
ATC	Air Traffic Control
ATCS	Advanced Train Control Systems
ATCSCC	Air Traffic Control System Command Center
ATHR	Auto thrust System
ATIS	Automatic Terminal Information Service
ATM	Air Transportation Management
ATN	Aeronautical Telecommunications Network
ATS	Automatic Throttle System
ATSC	Air Traffic Service Communications
AUG	August
AV	Avianca
AVAIL	Available
AVHRR	Advanced Very High-Resolution Radiometer
AWACS	Airborne Warning And Control System
AWAS	Automated Weather Advisory Station
AWIPS	Advanced Weather Interactive Processing System

B

B	At or Below (constrained altitude)
BALPA	British Air Line Pilots Association
BASIS	British Airways Safety Information System
BF	MarkAir
BIT(E)	Built-In-Test (Equipment)
BRG	Bearing
BRT	Brightness

C

C	Centigrade
CAA	Civil Aviation Authority (Great Britain)
CAB	Civil Aeronautics Board
CAAC	Civil Aviation Authority of China
CAS	Calibrated (Computed) Air Speed
CASE	Computer Aided Software Engineering
CAT	Clear Air Turbulence
CAT	Category
Cat II	A Cat II approach
CBT	Computer Based Training
CDI	Course Deviation Indicator
CDU	Control display unit (Interface to the FMS)
CDTI	Cockpit Display of Traffic Information
CENA	Centred' Études de la Navigation Aérienne
CFIT	Controlled Flight Into Terrain
CG	Center of Gravity
CGS	Centimeter-gram-second
CI	Cost Index
CI	China Airlines
CIT	Compressor Inlet Temperature
CLB	Climb Detent of the Thrust Levers
CLR	Clear
CMC	Central Maintenance Computer
CNS	Communications Navigations and Surveillance
CO	Continental
COM	Cockpit Operating Manual
CON	Continuous
CO ROUTE	Company Route (also CO RTE)
COTR	Contracting Officer's Technical Representative
COTS	Commercial Off The Shelf
CP	Control Panel
CPCS	Cabin Pressure Control System
CPDLC	Controller Pilot Datalink Communications
CPU	Central Processing Unit
CRC	Cyclic Redundancy Check
CRM	Cockpit Resource Management
CRM	Crew Research Management
CRS	Course
CRT	Cathode Ray Tube
CRZ	Cruise
CSD	Constant Speed Drive
CTA	Controlled-Time of Arrival
CTA	Control Area (ICAO Term)
CTAS	Center TRACON Automation System
CTC	Centralized Train Central
CTR	Center
CTR	Civil Tilt Rotor
CTRL	Control
CVSRF	Crew-Vehicle Simulation Research Facility (NASA Ames)
CWS	Control Wheel Steering

D

D	Derated
DA	Descent Advisor
DBS	Direct Broadcast Satellite
DE-TO PR	Derated Takeoff Engine Pressure Ratio
D-TO NI	Derated Takeoff Engine Fan Speed
DADC	Digital Air Data Computer
DATALINK	Digitized Information Transfer (air/ground)
DC	Direct Current Electricity
D/D	Drift Down
DEC	December
DEC	Digital Equipment Corporation
DECR	Decrement
DEL	Delete
DEP	Departure
DES	Descent
DEST	Destination
DEV	Deviation
DFDAU	Digital Flight Data Acquisition Unit
DFDR	Digital Flight Data Recorder
DFGS/C	Digital Flight Guidance System/Computer
DFW	Dallas Fort Worth International Airport
DGPS	Differential GPS
DH	Decision Height
DIR	Direct
DIR/INTC	Direct Intercept
DIS	Distance
DISCR	Discrepancy
DIST	Distance
DL	Delta
DLP	Data Link Processor
DLR	German Aerospace Research Establishment
DME	Distance Measuring Equipment
DMU	Data Management Unit
DNTKFX	DownTrack Fix
DOT	Department of Transportation
DOD	Department of Defense
DRU	Data Retrival Unit
DSPY	Display (annunciation on CDU)
DTG	Distance-to-go

E

E	East
EADI	Electronic Attitude Director Indicator
EAS	Equivalent Airspeed
ECAM	Electronic Centralized Aircraft Monitor
ECON	Economy (minimum cost speed schedule)
ECS	Environmental Control System
E/D	End-of-Descent
EDF	Electricité de France
EEC	Electronic Engine Control
EFC	Expected Further Clearance
EFIS	Electronic Flight Instrument System
EGT	Exhaust Gas Temperature
EHSI	Electronic Horizontal Situation Indicator
EICAS	Engine Indicating Crew Alerting System
EIU	Electronic Interface Unit
ELT	Emergency Locator Transmitter
EMP	Electromagnetic Pulse
EMS	Emergency Medical Services
ENG	Engine
E/O	Engine-Out
EPR	Engine Pressure Ratio
EPROM	Erasable Programmable Read-Only Memory
EST	Estimated
ETA	EstimatedTime of Arrival
ETX	End of Transmission
EXEC	Execute

F

F	Fahrenheit
FA	Final Approach
FAA	Federal Aviation Administration

FADEC	Full Authority Digital Engine Control
FAIL FMC	Fail Failure The inability of a system
FAF	Final Approach Fix
FANS	Future Air Navigation Systems
FAR	Federal Aviation Regulations (United States)
FAR	Federal Acquisition Regulation
FAST	Final Approach Spacing Tool
FBO	Fixed Based Operator
FCC	Flight Control Computer
FCU	Flight Control Unit
F/D or(FD)	Flight Director
FDAMS	Flight Data Acquisition and Management System
FDC	Flight Data Company
FDR	Flight Data Recorder
FEATS	Future European Air Traffic System
FEB	February
FF	Fuel Flow
FGS/C	Flight Guidance System/Computer
FIR	Flight Information Region
Fix	Position in space usually on aircraft's flight plan
FL	Flight Level (FL 310 For example)
FLCH	Flight Level Change
FLIDRAS	Flight Data Replay and Analysis System
FLT	Flight
FMA	Flight Mode Annunciator
FMC	Flight Management Computer (also FMCS - FMC System)
FMGC	Flight Management Guidance Computer
FMGS	Flight Management Guidance System
FMS	Flight Management System
FO	First officer
FOQA	Flight Operations Quality Assurance
FPA	Flight Path Angle
FPA	Focal Plane Array
FPM	Feet Per Minute
FQIS	Fuel Quantity Indicating System
FR	From
FRA	Flap Retraction Altitude
FRA	Federal Railroad Administration
FREQ	Frequency
FSF	Flight Safety Foundation
FT	Feet

G

GA	Go-Around
GA	General Aviation
GAR	Go-Around
GCA	Ground-controlled Approach
GDLP	Ground Data Link Processor
GHz	Gigahertz
GMT	Greenwich MeanTime
GNSS	Global Navigation Satellite System
GPS	Global Positioning System
GPWS	Ground Proximity Warning System
GRAF	Ground Replay and Analysis Facility
GRP	Geographical Reference Points
GS	Glide Slope
GS	Ground Speed
G/S	Glideslope
GSFC	Goddard Space Flight Center
GW	Gross Weight

H

HAC	Hughes Aircraft Co
HAI	Helicopter Association International
HBARO	Barometric Altitude
HDG	Heading
HDG	SEL Heading Select
HDOT	Inertial Vertical Speed
HE	Altitude Error
HF	High Frequency
HI	High
HIRS	High-Resolution Infrared Sounder
HP	Holding Pattern
HPRES	Pressure Altitude
HSI	Horizontal Situation Indicator
HUD	Head-Up Display

I

IA	Inspection Authorization
IAOA	Indicated Angle-of-Attack
IAS	Indicated Airspeed
ICAAS	Integrated Control in Avionics for Air Superiority
ICAO	International Civil Aviation Organization
ID	Identifier
IDENT	Identification
IEPR	Integrated Engine Pressure Ratio
IF	Intermediate Frequency
IFR	Instrument Flight Rules
IFRB	International Frequency Registration Board
IGFET	Insulated Gate Field Effect Transistor
ILS	Instrument Landing System
IMC	Instrument Meteorological Conditions
INBD	Inbound
INFO	Information
in.hg.	inches of mercury
INIT	Initialization
INR	Image Navigation and Registration
INS	Inertial Navigation System
INTC	Intercept
IPT	Integrated Product Team
IRS	Inertial Reference System
IRU	Inertial Reference Unit
ISA	International Standard Atmosphere
ISO	International Standards Organization
ITU	International Telecommunications Union

J

JAL	Japan Air Lines
JAN	January
JAR	Joint Airworthiness Regulations
JATO	Jet Assisted Takeoff
JL	Japan Air Lines
JSRA	Joint Sponsored Research Agreement
JUL	July
JUN	June

K

KG	Kilogram
kHz	kilohertz
KLM	Royal Dutch Airlines
km	Kilometer
KT	(kts) Knots
kW	Kilowatt

L

L	Left
LAT	Latitude
LAX	Identifier for Los Angeles
LCN	Local Communications Network.
LDGPS	Local DGPS
LFR	Low-frequency Radio Range
LIM	Limit
LMM	Compass locator at the middle marker
LNAV	Lateral Navigation
LO	Low
LOC	Localizer Beam
LOE	Line Oriented Evaluation
LOFT	Line Oriented Flight Training
LOM	Compass Locator at the Outer Marker
LON	Longitude
LORAN	Long Range Navigation
LOS	Line-Oriented Simulation
LRC	Long Range Cruise
LRU	Line Replaceable Unit
LVL	CHG Level Change

M

M	Mach Number
M	Manual Tuned NAVAID
MAA	Maximum Authorized IFR Altitude
MAG	Magnetic
MAINT	Maintenance
MAN	Manual
MAP	Missed Approach
MAR	March
M/ASI	Mach/Airspeed Indicator
MAX	Maximum
MAX CLB	Maximum engine thrust for two-engine climb
MAX CRZ	Maximum engine thrust for two-engine cruise
MCA	Minimum Crossing Altitude
MCDU	Multipurpose Control Display Unit
MCP	Mode Control Panel (pilots' interface to the autoflight system)
MCT	Maximum Continuous Thrust
MCW	Modulated Continuous Wave
MDA	McDonnell-Douglas Aerospace
MDA	Minimum Descent Altitude
MDL	Multipurpose Data Link
MEA	Minimum Enroute Altitude
MEL	Minimum Equipment List
MIDAS	Man-Machine Integration Design and Analysis System
MIDAS	Multi-discipline Data Analysis System
MILSPEC	Military Specifications
MIN	Minutes
MIN	Minimum
MIT	Massachusetts Institute of Technology
MLA	ManeuverLimited Altitude
MLE	Landing Gear Extended Placard Mach Number
MLS	Microwave Landing System
MMO	Mach Max Operating
MN	Magnetic North
MOA	Memorandum of Agreement
MOCA	Minimum Obstruction Clearance Altitude
MOD	Modified/Modification
Mode	Type of secondary surveillance radar (SSR) equipment.
MODIS	Moderate-resolution Imaging Spectrometer
MRA	Minimum Reception Altitude
MSG	Message
MSL	Mean Sea Level
MTBF	Mean Time Between Failures
MU	Management Unit
MWP	Meteorological Weather Processor

N

N	North
NACA	National Advisory Committee for Aeronautics
NADIN	II National Airspace Data Interchange Network II
NAS	National Airspace System
NAS	National Aircraft Standard
NASA	National Aeronautics and Space Administration
N/A	Not Applicable
NATCA	National Air Traffic Controllers Association
NAV	Navigation
NAVAID	Navigational Aid
NBAA	National Business Aircraft Association
NGATM	New Generation Air Traffic Manager
ND	Navigation Display
NDB	Nondirectional Radio Beacon
NESDIS	National Environmental Satellite
NLM	Network Loadable Module
NLR	National Research Laboratory (The Netherlands)
NM	Nautical Mile
NMC	National Meteorological Center
NOAA	National Oceanic and Atmospheric Administration
NOTAM	Notice for Airman
NOV	November
NRP	National Route Program
NTSB	National Transportation Safety Board
NW	Northwest Airlines
NWS	National Weather Service
NI	Engine Revolutions per Minute (percent)

O

OAG	Official Airline Guide
OAT	Outside Air Temperature
OATS	Orbit and Attitude Tracking
OBTEX	Offboard Targeting Experiments
OCT	October
ODAPS	Operational OGE Data Acquisition and Patch Subsystem
OFST	Lateral Offset Active Light
OGE	Operational Ground Equipment
OIS	OGE Input Simulator
OO	SkyWest Airlines
OP	Operational
OPT	Optimum
O-QAR	Optical Quick Access Recorder
OSI	Open Sytem Interconnection
OTFP	Operational Traffic Flow Planning
OV	Overseas National Airways

P

P	Procedure-Required Tuned NAVAID
PA	Pan Am
PAR	Precision Approach Radar
PAWES	Performance Assessment and Workload Evaluation
PBD	Place Bearing/Distance (way point)
PD	Profile Descent
PDB	Performance Data Base
PDC	Pre Departure Clearance
PERF	Performance
PF	Pilot Flying
PFD	Primary Flight Display
PHARE	Program for Harmonized ATC Research in Europe
PHIBUF	Performance Buffet Limit
PHINOM	Nominal Bank Angle
PIREPS	Pilot Reports
PMS	Performance Management System
PND	Primary Navigation Display
PNF	Pilot Not Flying
POS	Position
POS	INIT Position Initialization
POS	REF Position Reference
PPI	Plan Position Indicator
PPOS	Present Position
PREV	Previous
PROC	Procedure
PROF	Profile
PROG	Progress Page on MCDU
PROV	Provisional
PS	Pacific Southwest Airways
PT	Total Pressure
PTH	Path
PVD	Plan View Display

Q

QAR	Quick Access Recorder
QNH	Quantity
QRH	The barometric pressure as reported by a particular station
QTY	Quantity
QUAD	Quadrant

R

R	Right
R	Route Tuned NAVAID
RAD	Radial
RAD	Radio
RAPS	Recovery Access Presentation System
RASCAL	Rotorcraft Air Crew Systems Concepts Airborne Laboratory
RCP	Radio Control Panel
R/C	Rate of Climb
RDP	Radar Data Processing (system)
REF	Reference

REQ	Required/Requirement
REQ	Request
RESTR	Restriction
RESYNCING	Resynchronizing
rf	radio frequency
RMPs	Radio Management Panels
RNAV	Area Navigation
RNP	Required Navigation Performance
ROUTER	ATN network layer
RTA	Required Time of Arrival
RTCA	Radio Technical Committee on Aeronautics
RTE	Route
RVR	Runway Visual Range
RW	Runway

S

S	South
SA	Situation Awareness
SAS	Scandinavian Airlines System
SAT	Static Air Temperature
SATCOM	Satellite Communications
SBIR	Small Business Innovative Research
S/C	Step Climb
SEA/TAC	Seattle/Tacoma International Airport
SEL	Selected
SEP	September
SESMA	Special Event Search and Master Analysis
SID	Standard Instrument Departure
SIGMET	Significant Meteorological Information
SITA	Société Internationale Télécommunique Aéronautique
SO	Southern Airways
SOP	Standard Operating Procedure
SOPA	Standard Operating Procedure Amplified
SP	Space
b	Speed Mode
SPS	Sensor Processing Subsystem
SQL	Structured Query Language
SRP	Selected Reference Point
SSFDR	Solid-State Flight Data Recorder
SSM	Sign Status Matrix
STAB	Stabilizer
STAR	Standard Terminal Arrival Route
STEPCLB	StepClimb
STOL	Short Takeoff and Landing
STTR	Small Business Technology Transfer Resources
SUA	Special Use Airspace
SWAP	Severe Weather Avoidance Program

T

TACAN	Tactical Air Navigation
TACH	Tachometer
TAI	Thermal Anti-Ice
TAP	Terminal Area Productivity
TAS	True Airspeed
TAT	Total AirTemperature
TATCA	Terminal Air Traffic Control Automaiton
TBD	To Be Determined
TBO	Time between Overhauls
TBS	To Be Specified
TCA	Terminal Control Area
TCAS	Traffic Alert & Collision Avoidance System
T/C or (TOC)	Top-of-Climb
T/D or (TOD)	Top-of-Descent
TDWR	Terminal Doppler Weather Radar
TEMP	Temperature
TFM	Traffic Flow Management
TGT	Target
THDG	True Heading
THR	Thrust
THR HOLD	Throttle Hold
TI	Texas International
TIAS	True Indicated Airspeed
TKE	TrackAngle Error
TMA	Traffic Management Advisor
TMC	Thrust Management Computer

TMF	Thrust Management Function		VSCS	Voice Switching and Control System
TMU	Traffic Management Unit		VSI	Stalling Speed in a Specified Flight Configuration
TN	True North		VSO	Stalling Speed in the Landing Configuration
T/O / (TO)	Takeoff		VSTOL	Vertical or Short Takeoff and Landing
TOD	Top of Descent		VTK	Vertical Track Distance
TO EPR	Takeoff Engine Pressure Ratio		VTOL	Vertical Takeoff and Landing
TO NI	Takeoff Engine Fan Speed		V/TRK	Vertical Track
TOGA	Takeoff/Go-Around		VTR	Variable Takeoff Rating
TOT	Total		VU	Utility Speed
TRA	Thrust Reduction Altitude		VX	Speed for Best Angle of Climb
TRACON	Terminal Radar Approach Control Facility.		VY	Speed for Best Rate of Climb
TRANS	Transition		V1	Critical Engine Failure Velocity (Takeoff Speed)
TRK	Track (to a NAVAID)		V2	Takeoff Climb Velocity
TRU	True			
TSRV	Transport Systems Research Facility			
TT	Total Temperature			
TURB	Turbulence			

U

UA	United
UHF	Ultra-high Frequency
US	USAir
USAF	United States Air Force

V

V	Velocity
VA	Heading to an Altitude
VA	Design Manoeuvring Speed
VAR	Variation
VAR	Volt-amps Reactive
VAR	Visual-aural Radio Range
VASI	Visual Approach Slope Indicator
VBF(LO)	Flaps up minimum buffet speed
VBFNG(HI)	High speed CAS at N g's to buffet onset
VBFNG(LO)	Low speed CAS at N g's to buffet onset
VCMAX	Active Maximum Control Speed
VCMIN	Active Minimum Control Speed
VC	Design Cruising Speed
VD	Design Diving Speed
VD	Heading to a DME distance
VF	Design Flap Speed
VFE	Flaps Extended Placard Speed
VFR	Visual Flight Rules
VFXR(R)	Flap Retraction Speed
VFXR(X)	Flap Extension Speed
VG	Ground Velocity
VGND	Ground Velocity
VH	Maximum Level-flight Speed with Continuous Power
VHF	Very-high Frequency
VHRR	Very High-Resolution Radiometer
VISSR	Visible Infrared Spin Scan Radiometer
VI	Heading to a course intercept
VIs	Lowest Selectable Airspeed
VLE	Landing Gear Extended Placard Airspeed
VLO	Maximum Landing Gear of Operating Speed
VLOF	Lift-off Speed
VM	Heading to a manual termination
VMC	Visual Meteorological Conditions
VMC	Minimum Control Speed with Critical Engine Out
VM(LO)	Minimum Maneuver Speed
VMAX	Basic Clean Aircraft Maximum CAS
VMIN	Basic Clean Aircraft Minimum CAS
VMO	Velocity Max Operating
VNAV	Vertical Navigation
VNE	Never-exceed Speed
VNO	Maximum Structural Cruising Speed
VOM	Volt-ohm-milliammeter
VOR	VHF OmniRange Navigatgion System
VORTAC	VHF Omni Range Radio/Tactical Air Navagation
VPATH	Vertical Path
VR	Heading to a radial
VR	Takeoff Rotation Velocity
VREF	Reference Velocity
VS	Design Speed for Maximum Gust Intensity
V/S	Vertical Speed/Vertical

W

W	West
WAAS	Wide Area Augmentation System
Waypoint	Position in space usually on aircraft's flight plan
WBC	Weight and Balance Computer
WINDR	Wind Direction
WINDMG	Wind Magnitude
WPT	Way point
W/MOD	With Modification of Vertical Profile
WMSC	Weather Message Switching Center
WMSCR	Weather Message Switching Center Replacement
WO	World Airways
W/STEP	With Step Change in Altitude
WT	Weight
WX	Weather
WXR	Weather Radar

X

XTK	Crosstrack (cross track error)
XY	Ryan Air

Z

Z	Zulu (GMTtime)
ZFW	Zero Fuel Weight
ZNY	New York Air Route Traffic Control Center

Morse Code

A	.-	N	-.
B	-...	O	---
C	-.-.	P	.--.
D	-..	Q	--.-
E	.	R	.-.
F	..-.	S	...
G	--.	T	-
H	....	U	..-
I	..	V	...-
J	.---	W	.--
K	-.-	X	-..-
L	.-..	Y	-.--
M	--	Z	--..

0	-----	Full-stop	.-.-.-
1	.----	Comma	--..--
2	..---	Colon	---...
3	...--	Question (?)	..--..
4	....-	Apostrophe	.----.
5	.....	Hyphen	-....-
6	-....	Fraction bar	-..-.
7	--...	Brackets ()	-.--.-
8	---..	Quotation	.-..-.
9	----.		

Brain Teaser
Answers

Q1) Turn Right onto 011 and you will intercept the radial at a 45 degree angle.

Q2) This is a Surveillance Radar Approach and is a procedure in which the air traffic controller monitors your position on the radar and guides you down to the runway. You would be given headings to steer and the height you should be at that moment.

Q3) Carburettor heat is used to de-ice the carburettor and the screen heat is used to de-ice the windshield.

Q4) Above Glasgow airport. You could not receive the radio signal if you were on the ground. The beacons and their idents are St Abbs SAB and Turnberry TRN.

Q5) Heading 242 will put you back on track by the half way point and heading 249 will then keep you on track to the end of that sector. Dealing with the second heading first. As you have been blown off track by 7 degrees right, then flying 7 degrees left would have kept you on track. This is calculated as 256 minus 7 degrees of unplanned drift. You have been flying 7 degrees right of the best heading for a quarter of the sector, so flying 7 degrees left of the newly calculated corrected heading will take you back on track in the next quarter of the section. Hence 249 minus a further 7 degrees will do the trick. The sums are easy because each of the subsections is the same length.

Q6) No, you will not even get close over this long distance. The shortest distance between any two points on the roughly spherical world is a "great circle". If you want to fly along a great circle from London to New York, you will have to constantly change your heading. By the time you reached New York, your heading would have to be about 231 True.

Q7) No, you are flying a curved rhumb line, which is a line of constant heading. To fly a straight line you would have to fly a great circle. Over this short distance the difference is negligible.

Q8) The colours give a rapid identification of certain key speeds. The green arc is the normal manoeuvring speed range and the bottom of this arc is the normal flaps up stall speed VS1. The white arc stretches from the normal flap limiting speed VFE to the stall speed in the landing configuration with flaps and undercarriage down VSO. The yellow arc is the maximum structural cruising speed up to the VNE or Velocity Never Exceed.

Q9) These are the de-icing boots. When flying in icing conditions, they are periodically inflated and the ice building up on the front of the wing is broken off.

Q10) A radial engine has the cylinders fixed immovably in radial positions around the propeller shaft like the spokes of a stationary wheel. There are usually an odd number of cylinders between seven and eleven. A rotary engine looks much the same when stopped, but when it starts up, the central shaft stays still and all the cylinders spin round with the propeller.

Q11) Despite the acronym not being a perfect match, LITAS is Low Intensity Two Colour Approach Slope System. It is a night landing system that is similar to VASIS but with only two lights. You can try them out at Sywell aerodrome.
12 Turn your transponder to "standby" and then set the

numbers 3662 on it. Turn the transponder back to the "on" position and press the ident button. Now instead of the ATC radar showing you as just a small dot, it shows 3662 as well. Identing will make your number stand out from the others for a short period so you can be quickly identified by the controller.

Q13) The asterisk is used to indicate that the ILS is not available at all times. It can be used for other radio frequencies too.

Q14) A rate one turn is one which will turn you through a full 360 degree circle in two minutes. You would be turning at a rate of three degrees per second. It is the standard instrument flying turn.

Q15) The category of a plane is based on 1.3 times the stall speed. The boundaries are up to 90 knots for category A, 120 knots for B, 140 knots for C, 165 knots for D, which leaves anything 166 and above in category E. These categories are used to give slightly different approaches for planes that need to fly the approach at different speeds. Despite this, if you flew an instrument approach in a Cessna 172 at 100 knots, you would use the minima for category B.

Q16) It is normal to start the approach with the QNH set on the altimeter. Landing is normally done with the QFE set. The change over takes place at an unspecified point. To allow for this, the approach plate will show both the heights above sea level in a bold font, and the heights above the aerodrome which are usually shown in brackets or a lighter font.

Q17) A pilot flying VFR approaching an airport to land may be asked to orbit until some other plane gets out of the way. The pilot would simply fly round in circles over a fixed position until cleared to move on. A holding pattern is a much more exacting procedure flown by IFR traffic. It is normally a standard 4 minute race track pattern, which is started by flying over a specific point (usually a beacon) in a specific direction. This is followed by a 180 degree rate one turn, then one minute in the opposite direction and another 180 degree turn with one minute back to the starting point, to pass over it again in the original direction. This all sounds easy until you try it, especially in high winds.

Q18) The wind is 45 degrees across your track, so the crosswind and headwind components are both 14 knots. At 120 knots you would have to allow half of this (7 degrees) for the crosswind when flying inbound towards the beacon. Multiply this drift by three to get your drift for the outbound leg. To handle the headwind, add or subtract one second per knot of wind speed to the time on the two straight legs. So, at the beacon you would start your stop watch and turn left and then outbound on 246 for 46 seconds, then turn left again and fly inbound 038 for 74 seconds. You would doubtless have to make adjustments next time round when you see how this works out, but this would be a good starting point.

Q19) When flying at high angles of attack, approaching the stall, the wedges will stall the inner part of the wing first. This creates a buffeting vibration that can be felt by the pilot giving warning that the full stall is imminent. Because the inboard part of the wing stalls first, the outboard part of the wing retains some aileron roll control during the initial stages of the stall.

Q20) The Wright Flyers and other early aircraft were controlled in roll this way. They did not have ailerons but twisted the whole wings instead. Needless to say the roll response was rather poor.

Q21) You would roll out of the turn at an indicated heading of 150 degrees. The best acronym for this is UNOS, which stands for Undershoot North, Overshoot South. You are turning onto South, so you need to overshoot by 30 degrees.

Q22) The landing 747 would have created a significant amount of wake turbulence. This is a rotating vortex (or two) that could easily flip a light aircraft upside down. The recommendation, which only the foolhardy would ignore, is to be at least 8 miles and 4 minutes behind the 747. You could follow a higher approach and land further down the runway, but why take the risk.

Q23) Lift enables a plane to fly, and the amount of lift available to any plane is determined by the relative speed of the wind across the wings. It is the airspeed that counts. Imagine a plane needs 60 knots to fly and the wind speed is 20 knots. Taking off into wind, the plane only has to accelerate to a ground speed of 40 knots before it can become airborne. Taking off downwind the plane will have to accelerate to 80 knots before it reaches an airspeed of 60 knots, taking far more runway and creating far more grey hairs on the pilot.

Q24) This is a ground signal that the take off and landing direction is parallel to the main stem of the "T" and towards the cross piece. So as the letter "T" is written on this page, you would land flying from the bottom of the page towards the top.

Q25) The marshal is instructing you to stop. The next signal is likely to be for you to cut your engines.

Q26) The pressure reduces in a depression and that's why the are often called "Lows". As you fly towards the low pressure you would need to adjust your altimeter to use the updated QNH. If you don't you will get closer to the ground, which is not what you want to do as you fly into the associated cloud. This problem is solved by using a regional QNH.

Q27) You have 3 options for working this out. The first is to imagine where the ADF needle superimposed on top of your Direction indicator would be. The second is to turn the rotating compass card on the ADF until it has your heading of 315 at the top and the needle will show the QDM. The last option is to add the two figures together (315 + 235 = 550 degrees) and if necessary subtract a full circle of 360 degrees (550 - 360 = 190) and the QDM is 190.

Q28) During a QGH letdown, the air traffic controller uses direction finding equipment (VDF Very high frequency Direction Finding station) to monitor where the plane is in the sky. The VDF shows the direction of radio transmissions and therefore the pilot must transmit occasionally for the direction finding equipment to follow him. The controller uses the VDF and a stopwatch to guide the plane round the aerodrome approach procedure.

Q29) There are several mnemonics for learning Morse code. C is Dash Dot Dash Dot, or Daar Dit Daar Dit, which sounds a bit like "Charley Charley". Similarly Q is Dash Dash Dot Dash, which has the same rhythm as the familiar wedding march.

Q30) Yes it could be overcast. CAVOK means that the visibility is 10 kilometres or more, there is no precipitation, thunderstorm, cumulonimbus clouds, fog or low drifting snow. It also means that there is no cloud below 5000 feet or the minimum sector altitude. Hence it could actually be 10 k visibility and overcast at 5500 feet blowing a gale.

Q31) Looking from above the plane has to turn right. To do this, the plane has to roll right, so push the stick to your right (you do this to roll right whatever position the plane is in). You also want the rudder to move to the right as viewed from above, which is the left as viewed from below, so press the left rudder with your left foot.

Q32) The accelerometer shows how many gee the plane's wing is pulling. In a steady climb of 45 degrees this will be 0.7 (Cosine 45, sorry about the maths). As you are upside down, the gauge will show minus 0.7 gee. Give yourself an extra pat on the back if you said that the plane is probably decelerating and will be pulling a little less than 0.7 gee.

Q33) The light appears stationary in the sky, which means that if it is getting closer to you, it will eventually crash into you. Unfortunately at night you probably cannot tell if it is getting closer or not. The other plane is on your right, so it has the right of way, therefore it would be sensible to turn right to pass behind it.

Q34) The correct sequence is to level off and let the speed build up to your cruise speed. Then you can reduce the throttle, set the correct RPM and then set the mixture. When everything has stabilised, trim the plane for level flight. Note that many pilots would trim a little after levelling off to make things easier for themselves, despite this not being the official sequence.

Q35) Close the throttle and centralise the ailerons holding the stick back, then apply full right rudder, wait a few seconds and then steadily move the joystick fully forwards. The spin will stop and you will enter a steep dive. Let the speed build up to a decent flying speed and then pull out of the dive. There is an alternative method, which works on some planes, of simply closing the throttle and letting go of everything.

Q36) On some planes it is particularly important that the ailerons are exactly in the centre when trying to get out of a spin. The idea is that the joystick is pushed forwards until it reaches the white line, which is in exactly the right position.

Q37) Three degrees is the most common, but it can be much steeper than this. London City ILS is 5.5 degrees.

Q38) I have received all of your last transmission. Note that it does not mean that you will do anything about it, or even comply.

Q39) The transition altitude at an aerodrome is the altitude below which a pilot should use the aerodrome QNH (or QFE) on his altimeter. The transition level is the Flight Level above which the pilot should use the QNE (1013 mBars). This is because all en-route instrument flying is generally done on the standard setting of 1013 millibars. Doing this means that everyone is using the same altimeter setting and everybody's relative height is obvious. During landing and take off, when you are close to the ground, it is better to know your actual height. When using the QNE, the reading on the altimeter divided by 100 is the "flight level".

Q40) These are the regional altimeter QNH settings for the Tyne ASR (Altimeter Setting Region) and the Barnsley ASR. Teesside Airport is almost on the boundary between these two ASR, so ATC has given you both. The pressures are the lowest predicted QNH anywhere in the ASR and it is normally updated every hour. You would use the appropriate one for your VFR flying.

Q41) No, you should pick a field or somewhere straight in front of you that is suitable to land. Turning back for the field can be done if you are extremely good, prepared to fly in riskier than normal flight modes and willing to bet your life on getting it right. It's the kind of thing you can do in a simulator where you can just restart when you don't make it.

Q42) The trick is side slipping. Stay at your approach speed, push the stick one way and press the opposite rudder just enough to prevent the plane turning. The plane will yaw a bit as you initiate the side slip, but once the side slip is established, the nose should stay pointing in a constant direction. For some reason this always seems much easier on real aircraft than on PC simulators. Remember to turn off any automatic coordination if you want to try it.

Q43) The circuits are usually left hand. This is so that the captain, who normally sits on the left hand side, has a good view of the runway most of the time.

Q44) You will have to press with your left leg. You always "Tread on the ball" to centralise it. There is also a mnemonic, "Dead engine, dead leg".

Q45) Yes, you are in the optimum conditions for carburettor icing. You would apply the carb heat for approximately 30 seconds every ten minutes or so.

Q46) It is a shorthand way of saying "change frequency" and would be used as follows "Golf Tango Mike Alpha Golf request QSY to Southend Approach on 128 decimal 95"

Q47) Failure of the suction pump will affect the instruments that are powered by suction. These are normally (but not always) the Attitude Indicator and the Directional Gyro. Following a suction pump failure, these will take some time to wind down, but they will ultimately become useless. Monitoring the suction gauge may give you early warning, enabling you to easily move to limited panel flying.

Q48) The gyroscopic instruments in some planes can be seriously disrupted by aerobatic manoeuvres which take the instruments past their normal limits of movement. Many of them will be totally confused by rolling whilst flying vertically upwards or downwards. This disruption is known as "toppling" and the results in erratic behaviour of the instruments. To prevent this happening, some instruments can be fixed in position, or "caged", before beginning the manoeuvres and then released afterwards.

Q49) The first thing to do is to open (increase) the throttle a little, or a lot if you are far too low. You may need to pull back on the stick a little, but this is a secondary action.

Purists state that on final approach you should control your speed with the joystick and your height with the throttle.

Q50) You would pull two gee in the properly executed 60 degree banked turn. These steep turns often require a good boot-full of rudder to balance them correctly. You would be amazed how often such turns are held at the correct angle, but unbalanced with nothing like two gee pulled.

Q51) Both instruments tell you the rate at which you are turning, however the aeroplane symbol on the turn indicator also responds to roll. This means that the turn indicator needle only registers once the plane has started to change heading, whereas the turn coordinator registers as soon as you start to roll into a turn. The turn coordinator is therefore more intuitive to use.

Q52) When a reversible pitch propeller is turned so that it produces zero or reverse thrust, it is said to be in its beta range. There is sometimes a beta for taxi range that gives a very small amount of thrust.

Crossword Answers

Across

1. FMC
5. Tower
7. Throttle
8. Yoke
9. Taxi
12. Navigation
13. Flaps
15. Pushback
17. Refuel

Down

2. Captain
3. Terminal
4. ICAO
6. Steward
10. Aircraft
11. Airbus
14. ATC
16. Bleriot
19. ILS

Word Search Answers

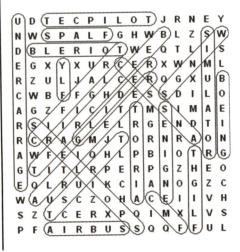

General Index

General Index

General Index

The Good Flight Simmer's Guide